The Art of Political Power

A Translation and Commentary of
The Book of Master Guan (The Guan Tzu)

Guanzi, Bill Bodri

Top Shape Publishing LLC
1135 Terminal Way Suite 209
Reno, NV 89502

ISBN: 978-1-7370320-3-8
Library of Congress Control Number: 2023938575

DEDICATION

In admiration of George Marshall, Lee Kuan Yew, Zhuge Liang, Cyrus the Great and historian Rufus Fears. Each of these men gave us useful models of a statesman.

CONTENTS

ACKNOWLEDGMENTS

My thanks go out to the many individuals who assisted in preparing and reviewing the translations for this text including Douglas Wile, William Brown, Mathew Tang, and Zhang Li-Chu. All these individuals prepared versions of the final translation or reviewed the final version of the text. Of these individuals, I am most indebted to Douglas Wile for editing the final passages. Without his input, the text wouldn't possess anything near the polish it has today.

Paul Wong, Margaret Yuan, Hung Jen, Lee Shu Mei, Sami Kuo, and Paul Rath also have my sincerest thanks for their assistance in helping me to interpret various portions of this work, as well as Martial Gabin and Marshall Adair for their most helpful comments.

I must also pay my respects to the previous translations of the *Guanzi* produced by Lewis Maverick and W. Allyn Rickett, which served as invaluable references when trying to interpret this material. If you want to further research Guan Tzu's ideas and his methodologies or peruse his other writings then these translators have produced invaluable guides and they are the first place to start.

Finally, the greatest of thanks goes to Nan Huai-Chin, who first introduced me to Guan Tzu when I asked him which sage's writings could best help America today. Nan Huai-Chin told me:

> There are many Chinese sages who have written on the problems of humanity, and each has captured a portion of the larger picture and spoken of what was relevant for their time. But the choice of best political philosopher and geo-strategist for our times should end up with Guan Tzu and no other. Of the various Chinese sages who have spoken in the fields of economics, administration and politics, without doubt it is Guan Tzu whom Westerners should be studying if they wish to manage the problems of the world today.

The help of these individuals is why you have this text today.

INTRODUCTION

Guan Zhong (ca. 720-645 B.C.E.), who served as the chancellor or chief minister of the state of Qi during the Zhou dynasty, is considered one of ancient China's great sages. He is also known as Guan Zhong, Guan Tzu, Guanzi, Master Guan or Kuan Tzu. Sun Tzu is known for the tactics of warfare but Master Guan (is known for political and geopolitical strategy at a level far transcending Sun Tzu, including conquest by economic warfare. Guan Tzu (Guanzi) is one of the five great prime ministers of China whose strategies are considered superior to Sun Tzu's since he would win battles without fighting and yet few know his name.

Master Guan's highly inventive diplomatic policies during a time of turmoil unified China's separate states through treaty rather than through warfare, and thus he personally saved the country's imperial system from collapse. Guan Zhong's highly unique administrative, economic and modernization policies also helped his state of Qi become the most powerful state within China. Because of his great political achievements, as well as his legendary ability to turn his lord's mistakes into advantages, his policies and writings have subsequently served as a guide for political strategy and power politics for countless Chinese generations. He was one of the first people in history, for instance, to combine Keynesian and monetarist policies to run a state and conquer other countries. He was not Machiavellian because he stressed rule by virtue and benevolent means.

Guan Zhong, whom posterity has rewarded with the title of Master Guan, was born to a poor family in Ying-shang, a northwestern district of China's present Anhui province. While young, he and his good friend Bao Shuya were partners in a trading business which dealt with China's various feudal states. In their partnership Guan Zhong always took the larger share of the profits, which caused Bao Shuya's servants to complain, but Bao Shuya always defended Guan Zhong saying, "Guan Zhong isn't greedy. I let him take most of the profits because his family is poor."

At another time, Guan Zhong served as a low level officer in a military expedition and stayed in the rear when his army attacked. On the trip homewards, however, he returned at the very head of the column of troops. When some laughed at him and called him a coward, his friend Bao Shuya simply pointed out, "Guan Zhong's mother is old and alone, so he has to stay alive to take care of her. Actually, he isn't afraid to fight at all."

Bao Shuya often remarked, "A wise person can achieve nothing if he is born at the wrong time. I am sure that Guan Zhong will make a great figure if he is only given the chance." Three times Guan Zhong was made an official and then dismissed because no one recognized his talents, but Bao Shuya only looked at this and replied, "What a shame it is that Guan Zhong hasn't yet encountered a wise ruler." When Guan Zhong heard of his friend's remarks, he could only say, "My parents are the ones who gave me birth but it is Bao Shuya who really understands me!"

Because of their learning and ability, Guan Zhong, Bao Shuya and a third friend, Shao-hu, were each made an advisor to one of the three sons of Duke Xi, ruler of the Chinese state of Qi (which is now known as Shandong province). Guan Zhong was appointed advisor to Prince Jiu, Bao Shuya was appointed advisor to Prince Xiao Bai, and Shao-hu was made advisor to the duke's firstborn son. Of this trio, the friendship between Bao Shuya and Guan Zhong was particularly strong so that both agreed to recommend the other should their respective patrons assume the throne. In this way they could both enjoy high positions together if fate permitted it.

Duke Xi eventually named his firstborn as the successor to his throne, but this prince, who became Duke Xiang upon ascension, was an unpopular ruler because of extravagance and arrogance. He often quarreled with the feudal lords of China's other states and was equally disdainful of his homage duties toward the emperor. He was eventually killed in a palace scuffle organized by his own maternal uncle and two generals whom he had offended. Upon the outbreak of rebellion, Guan Zhong fled with Prince Jiu and his followers to the Chinese state of Lu, while Bao Shuya and Prince Xiaobai stationed themselves in the neighboring state of Ju.

The usurper to the throne, Gongsun Wuzi, did not win any support from his courtiers and soon was also assassinated. This gave Prince Jiu and Prince Xiaobai their chance to assume power. When the ruler of Lu, Duke Zhuang, heard that Qi's minister Gao Xi was welcoming Prince Xiaobai home as the new ruler he quickly sent Guan Zhong and some soldiers into Ju so that they could kill Prince Xiaobai before officially becoming ruler. During an ambush, Guan Zhong shot an arrow at Prince Xiaobai wounding him, and the arrow would have killed the prince had it not bounced off his metal belt buckle.

Thinking quickly, Bao Shuya told the prince to crouch down in their carriage feigning death, and the two continued racing onwards to Qi's capital where Prince Xiaobai was successfully installed as the new duke. Thinking the prince had been killed, however, Guan Zhong took his time to finish his journey.

Eventually, it was surprisingly discovered that Prince Xiaobai had indeed survived the arrow and ascended to the throne, assuming the coronation title of Duke Huan. Duke Zhuang of Lu was enraged at this turn of events, and sent an army into Qi in order to defeat the new duke and help Prince Jiu assume the rulership of Qi. However, the new duke won the conflict and demanded that his brother be put to death. The state of Lu, which was harboring Guan Zhong, had no choice but to throw him into prison as well.

Having become the uncontested ruler of Qi, Prince Xiaobai (the new Duke Huan) next decided to reward all those who had helped him win the dukedom, and asked his long term advisor Bao Shuya to become the state's chief minister. However, Bao Shuya declined saying he did not possess the requisite abilities for running the state! When Duke Huan next asked how he could then become supreme ruler of the entire realm, Bao Shuya surprisingly replied that the duke should employ the talents of Guan Zhong if he wished to become the supreme leader of China. Guan Zhong, he said, was the only man who could make the duke's dreams come true.

Naturally, the duke's immediate response was to reject such advice since Guan Zhong had previously tried to kill him in battle. It was only because the duke had been wearing a bronze belt buckle that his life had been saved. How could the duke forgive his would-be assassin, let alone adopt him as a trusted advisor? Knowing the duke's mind, Bao Shuya pointed out that at the time of the fighting Guan Zhong was simply serving his own master and doing what would be expected in this role. His attempt to kill the duke in order that that Prince Jiu could ascend the throne was quite natural and proper under these conditions. Bao Shuya also argued that Guan Zhong's own previous loyalty to Prince Jiu was also ample proof that he would be loyal to the new duke, who was now the lawful sovereign. Employing Guan Zhong was not only necessary, Bao Shuya argued, but using him was the correct course of action because it was to the duke's own benefit.

In a frank discussion with his ruler, Bao Shuya insisted that the duke should forget the past and employ Guan Zhong as an important minister because as a policy a ruler should always put aside his own personal feelings, including his dislikes and resentments, for the good of the state.

Inferiors, he reminded the duke, must always do that in regard to their superiors so a superior must sometimes do that for one of his inferiors. In this case, Bao Shuya argued that the duke should grant Guan Zhong amnesty and use him because only Guan Zhong had the genius to make the duke's hopes come true.

Furthermore, Bao Shuya also pointed out that if the new duke could both forgive and also grant a high position to someone like Guan Zhong who had offended him, such generous behavior would become known throughout the land. The magnanimous reputation thus earned would work greatly to the duke's advantage in attracting others with outstanding talents to Qi, and they could then also be used in building the state.

In time, perhaps because he was generous of heart or perhaps because he was simply shrewd enough to realize that Guan Zhong could help him, the duke concurred with Bao Shuya's advice and agreed to use Guan Zhong. Unfortunately, Guan Zhong was now held prisoner by the neighboring state of Lu, which quite recognized his great abilities and the danger to their state if he were freely employed as a senior minister. Such a talented individual could become a threat to Lu if his abilities were to be used by another state. Knowing they had to get around this suspicion, Duke Huan and Bao Shuya therefore came up with a clever strategy in order to regain Guan Zhong alive.

In this ruse, Duke Huan requested that Lu hand Guan Zhong over to Qi so that the duke might personally witness Guan Zhong's execution. Naturally the ruler of Lu was not totally naïve and harbored suspicions as to Duke Huan's true intentions. Nevertheless, he ordered that Guan Zhong be returned to the state of Qi. If Duke Huan did not really wish to employ Guan Zhong as a minister yet the state of Lu killed Guan Zhong without regard to Qi's request then a premature execution would only increase the enmity between the two states when their initial relations had already gotten off to a bad start. However, had Duke Zhuang of Lu known for certain that Guan Zhong would be used in the new government then he would certainly have killed his prisoner so that Qi could not avail themselves of Guan Zhong's talents. Weighing the risks, Duke Zhuang ordered that Guan Zhong be returned home.

And so, Duke Huan's ruse successfully tricked Lu's ruler into deciding to send Guan Zhong back to Qi. While Guan Zhong was being transported by prisoner's cart he remembered his good friend Bao Shuya and began to think of hope rather than death, for why else should Qi demand that he be returned alive rather than be executed? Surmising that the request was some ruse, Guan Zhong then began fearing that Duke Zhuang would suspect as much as well and that he might still send orders for his execution before reaching the border. Indicative of his strategic alertness, he therefore taught his military escort several marching songs so that the soldiers would quicken their pace in transporting him back to Qi. His hopes confirmed, Guan Zhong's life was spared upon arriving in Qi where he was met with great welcome by his good friend Bao Shuya.

Duke Huan subsequently forgave Guan Zhong for attacking him and on an auspicious day he held a grand ceremony for Guan Zhong's reception where he was made chief minister of the state. Guan Zhong quickly gained the complete confidence of the duke and held the position of chief minister for the next forty years until his death. In time, the duke so trusted Guan Zhong that he considered him as family, going so far as to call him grandfather. This can only be attributed to Guan Zhong's trustworthiness and his accomplishments, as well as to his personal self-cultivation.

There are several important lessons we can learn from this introductory story before moving on to the main text. Firstly, the duke showed that he was wise enough to recognize Guan Zhong's talents and sufficiently broad-minded—or opportunistic—to forgive his earlier trespass and employ him in an important position. Secondly, Bao Shuya, without ill will, was deferential to his friend's great genius and voluntarily assumed an inferior post in the government. Perhaps because of friendship, but definitely to his own credit, Bao Shuya admitted his own limitations and refused to occupy the country's highest post because he felt this was best for the country.

In both cases these men could be called "great" because they were able to forget their personal feelings for the good of the nation, which is what great leaders do. Such behavior is an admirable component of "statesmanship" and a necessary characteristic of those aspiring to lead men. This is the type of behavior which the *Guanzi* tries to teach.

In deciding which of the *Guanzi's* many chapters might be of the most use to a modern audience, I have selected Chapter 23 which teaches how to administer a state and how an individual might become the single leader of a country (its ruler), or a "Lord Protector," or even the supreme leader of the world. It is a chapter on political leadership and political or geopolitical supremacy.

The greater part of the *Guanzi* includes additional policy discussions on how to make a state supreme among rivals and how to use its power to unify a confederation of states which comprise either a single country or greater alliance. However, the advice on attaining a leadership position—with the relevant tactics and strategies one must employ—as well as the instructions for wisely managing a country (or other organization), seem the most appropriate for today.

The selection presented within was therefore chosen in order to help someone rise to the top of his or her organization—whether it be involved with political or business or other affairs—and details the methods necessary for achieving prosperity for the group. There are ample lessons for individuals who wish to become great political leaders, business executives who desire commercial success, the statesman who wishes to save his country, and for governments wishing to forge their own states into superpowers. This section of the *Guanzi* teaches one how to attain political leadership and become a supreme political leader. In short, the lessons are all about leadership strategy. You might call this chapter, "The Art of Political Leadership" or "Master Guan on Political Leadership."

Providing such teachings is always dangerous because they can come into the possession of the immoral who may be smart enough to use medicine as poison. The criminal-minded tend to be smarter than virtuous people because they must be more clever in order to survive. They are usually the first to use such strategies. Furthermore, the perverse and sinister are fond of using strategy to destroy or simply oppress others. Therefore, I always stress that you must also be careful as to whom you teach the laws of power. You must also be careful when you select your leaders. You should examine the signs, such as personal nicknames and past history, before you entrust someone with high office and the power that comes with it. Will a prospective leader act for selfishness and gain, or will he or she make the difficult decisions necessary for the country's welfare? These are just some of the concerns to consider.

In this modern age when most leadership positions are determined by elections rather than by conquest, you must recognize that the great leader is the one who will do everything for the sake of his country and not for the sake of his personal popularity, wealth, power or position. He may not necessarily be popular at all times or in all things, but popularity often has little to do with a leader actually saving a nation.

On this note, I often reflect upon the excellent response of the general Fabius Maximus, one of the few Roman emperors who were awarded the title of "Maximus." Fabius is famous for the strategy of having continually shadowed Hannibal's invading army without engaging it in battle despite the pressure and criticism of his countrymen. Despite a vociferous chorus of criticism and jeers Fabius stuck to his avoidance strategy and refused to fight with Hannibal knowing that the Roman troops were not strong enough to win, and in this manner he saved Rome from destruction. In response to his vociferous critics, Fabius never deviated from what he felt was best, giving the immortal response,

> I should be more faint-hearted than they make me, if, through
> fear of idle reproaches, I should abandon my own convictions.
> It is no inglorious thing to have fear for the safety of our

country, but to be turned from one's course by men's opinions, by blame, and by misrepresentation, shows a man unfit to hold an office such as this, which, by such conduct, he makes the slave of those whose errors it is his business to control.[1]

Such advice is precious, which is that to save a country a leader must often hold to an unpopular course despite the pain and public outrage it may bring. When the errors of decades are corrected at a stroke there is absolutely no way to avoid pain, protest and suffering. Taking bitter medicine is always an unpleasant affair, but if you refuse the medicine your illness cannot be cured. If political leaders can accept such pain and hold to a wise course despite rising opposition then they have learned the lesson of this book and can truly earn the title of a "Maximus" or "National Savior." These are our modern day equivalents to what the Chinese called a "Lord Protector."

---

[1] *Plutarch: Lives of the Noble Romans*, ed. by Edmund Fuller, (New York: Dell Publishing Company, Inc., 1976), pp. 71-72.

CHAPTER 1
With All My Faults Am I Still Qualified To Lead?

When Duke Huan, the ruler of Qi, and Guan Zhong first met, they talked for three days and nights without stop and discussed the duke's plans for becoming the supreme lord of China. The duke asked Guan Zhong, "I have three great vices. In spite of these, can I still rule the country?"

"I haven't heard anything about this," replied Guan Zhong. "What three vices are you talking about?"

"The first vice," said the duke, "is that I like to hunt and can disappear for days at a time on hunting expeditions. As a result, while I am away on these trips the business of the state stops because my officials, and the emissaries of the other lords, have no one to report to."

"Well, that is bad," replied Guan Zhong, "but it is not vital."

The duke continued, "My second vice is that I like drinking and entertainment at all hours of the day and night. When I'm being entertained the business of the state stops because my officials, and the emissaries of the other lords, have no one to report to."

"Yes, that is bad," replied Guan Zhong, "but it's not crucial."

"Well, my serious vice," the duke said next, "is that I'm preoccupied with women and sex. I love women so much that I haven't even married off my aunts and sisters."

"Yes, that is bad," replied Guan Zhong, "but it is not vital."

At this reply the duke became flushed with anger and shouted, "If these three things are okay, then tell me what isn't okay!"

Guan Zhong calmly replied, "When a ruler is too indecisive or too slow to take action, this is inappropriate behavior. When a ruler is too indecisive he will lose the support of the people. When he is too slow to act he will never undertake what is necessary in time.[2]

"If you want to lead a state then it is absolutely essential that you possess or that you cultivate these two characteristics. This is what is crucial, this is what is vital, this is the important thing. And because you are both decisive and quick to take action when necessary then despite these other vices you can indeed be successful in leading the country."

---

[2] As the *T'ai Kung's Six Secret Teachings* states, "If you see good but are dilatory, if the time for action arrives and you remain passive, if you know something is wrong but you sanction it—in these three the Tao stops."

CHAPTER 2
The Ancient Wisdom of Master Guan

Thus begins our tale of the Prime Minister Guan Zhong (respectfully titled Guan Tzu or Guanzi), which takes place during China's Zhou dynasty somewhere between 685 and 645 B.C.E. At this time in history, China's various feudal states were constantly engaged in wars of conquest and annexation where each state was vying to become foremost in the nation. Confucius therefore named these turbulent times the "Spring and Autumn" period to denote the successive rise and fall of these various states and their rulers.

It was the custom during this period for the feudal lords to seek the guidance of worthy ministers for governing their state affairs. Since the success or failure of any state rested squarely on the policies it followed a wise ruler was always on the lookout for worthy advisors who knew the art of making states peaceful and prosperous, and making their military forces powerful. This was the reason that Duke Huan, as ruler of Qi, eventually appointed Guan Zhong as his chief minister because in Guan Zhong he felt he had finally found the capability he was seeking.

At the time of our story the Chinese emperor had lost so much of his original power and prestige that the eight to ten feudal lords which ruled the smaller Chinese states making up the realm were able to treat him as a mere figurehead rather than as the supreme ruler of the land. Though the emperor was still considered sacrosanct and his position gave order and structure to society, his situation could be compared to that of the Mikados (emperors) of Japan who were rulers in name but subject to the control of the powerful Shoguns.

Various Roman Catholic popes had shared a similar fate during certain periods in European history. They were dependent upon sovereigns who both protected and manipulated the Church. Hence, the situation in China was not historically unique.

One could best summarize the overall situation by saying that in Guan Zhong's day the various Chinese states were neglecting their homage duties to the emperor. In their quest for power were presuming more and more imperial rights for themselves. Furthermore, each of the feudal lords (the "dukes" or "princes" controlling China's individual states) was engaged in a continuous struggle for dominance over his peers. China's national cohesion was not only threatened by these competitive struggles but by foreign danger as well, which was the threat of external Tartar invasions from the north.

When Duke Huan of Qi first met Guan Zhong he naturally inquired about his ideas on governing in this environment and asked Guan Zhong if he had any plans for how the duke might become the country's ultimate leader. So intent was the duke on accomplishing these goals that the two ended up discussing state matters for several days and nights without stop. They touched upon economics, diplomacy, social welfare, agriculture, increasing the wealth of the state, military defense, educational policy. In general they discussed Guan Zhong's overall strategy for making Qi into China's strongest and most prosperous state. Throughout these initial discussions and throughout Guan Zhong's long reign of office he emphasized that the duke or any other ruler should always act at the right moment and never bypass an opportunity because these are vitally necessary for being able to succeed in grand schemes.

To rule successfully, Guan Zhong furthermore stressed that a leader must address the call of the times. If a leader missed the opportunities that came to him and hesitated to act when the timing was proper—perhaps because he was unsure of himself or inflexible or ill-informed or unprepared—he would never be able to accomplish anything grand for his state. Opportunities typically come but once, so Guan Zhong warned that if a leader were too inflexible or unresponsive so as to let them pass by then those leaders would achieve little in their lifetime. You cannot accomplish everything by force so you must make use of opportunities to accomplish more with less effort.

What Guan Zhong also emphasized is the great importance one should place on making careful plans and preparations. Planning is absolutely essential for any leaders who wish to succeed in their activities because without careful planning the execution of strategy or tactics is bound to fail. President Dwight David "Ike" Eisenhower is famous for saying, "In preparing for battle I have always found that plans are useless, but planning is indispensible." As famous axiom runs, "Prior preparation and planning prevents piss poor performance."

Planning prepares you to be ready to act at an instant. For instance, all the preparation that a professional athlete puts into a contest—the game plans, analysis of certain moves, visualization rehearsals, special exercises and so on—is useless if he cannot put them into action when game time comes. In order to execute his policies Guan Zhong insisted on prior planning.

Guan Zhong also stressed that the leader of a nation must actively search for capable people who can develop as well as execute effective policies because leader's success requires making use of the finest talents available. No man is an island where he can do everything himself. To succeed in grand plans a leader needs to depend upon capable others. A ruler needs to avail himself of the greatest talents and resources available so he needs some means to attract and encourage the capable to step forward and assist him.

In Guan Zhong's view, rulers who neither made careful preparations for change beforehand, nor trusted capable ministers with power and authority, would never be able to institute beneficial policies for their country. The *I Ching* says in many places that inferior people should not be used to carry out important affairs. A good leader needs talented people with experience to devise effective practical measures to solve the problems of the state and many capable people to carry out those plans. There is always a top echelon of people who must bear the burden of carrying out government policies.

When a leader cannot recognize who is capable, or fails to recruit or use the capable talent he has available (perhaps appointing them to high positions but failing to trust them), he endangers his own cause. Hence the ability to recognize talent and prepare for change are some of Guan Zhong's essential requirements for governing a country.

This view seems quite disquieting when we compare it with the emphasis more common today. The national press and public in most countries now place more emphasis on their leader's sexual affairs than on their governing abilities and actual plans for the nation.[3] A nation's overriding concern should be whether a political leader actually possesses the requisite leadership abilities. Most leaders are not paragons of virtue but they certainly should be able to manage the state. When they can do that then countless other issues will fall into place.

---

[3] When King Xuan of the Chinese state of Qi confessed to Mencius that he was faulted with obsessive sexual longings and an extreme desire for wealth Mencius wisely replied (in a wonderful example of skillful means) that all people desire such things. The key for a ruler, he said, is to make it possible for the people to attain that which they also desire. If the ruler creates prosperity for the state then the people will not criticize him if the part he takes is a little more. As long as the behavior of a king does not hurt the feelings of the people then he can enjoy himself while keeping the nation at peace.

These various teachings and the majority of what we know about Guan Zhong or Master Guan come down to us from *The Book of Master Guan*, which is known in short as the *Guanzi*. The *Guanzi* is actually a compilation of some 24 smaller books with a total of 86 sections, 10 of which have been lost to posterity.

Many of the 76 sections which have been left to us cannot actually be attributed to Master Guan himself but are by many different authors over several successive centuries who appended Guan Zhong's name to their own writings in order to give them the air of authority. This was a common practice in ancient China where the tradition was to revere the ancient but not the modern, and so individuals would sometimes append some respected name onto their own writings in order that they might gain acceptability. Hence, a large portion of the *Guanzi* is undoubtedly by later authors who wished to add their own thinking to the master's original words.

The collection of essays in the *Guanzi* covers an extremely wide variety of governing matters although the majority of sections deal with questions of economics. However, the overall theme of the collection is to discuss various strategies and the effective tactics of power politics that will bring economic prosperity to a country, strengthen its military capabilities, and make its ruler the supreme leader of a region. In short, we can call these discussions "leadership strategies" that you can apply to not only running a country but also a business or any other organization.

A leader employing Master Guan's strategies might not be able to become the king of a country due to lacking the required heredity, but they could still become the strongest leader of a land and earn the title of the nation's "Lord Protector." This title referred to an individual who always held the country's best interests at heart and who always acted for the nation's overall benefit rather than for his own self-interest. The Lord Protector was in effect the preeminent leader of the country, someone who stood heads above the other lords because he was the most righteous and most powerful leader in the realm.

While Duke Huan could never become the legitimate emperor of China, Guan Zhong's policies effectively raised the duke to this position of preeminence. He became the first of the five hegemons in Chinese history who were especially powerful rulers during China's Spring and Autumn period. Duke Huan became China's first Lord Protector.

So effective were Guan Zhong's domestic and foreign policies that Qi became the most prosperous state in the realm and the duke won substantial influence over China's other feudal states, eventually presiding over three armed confederations and six peace conferences where previously there had been little cooperation between the states. By following Guan Zhong's policies and wearing the mantle of moral righteousness Duke Huan became the natural leader of these confederations because the other feudal lords, with their ministers and subjects, viewed his actions as unselfish because they were taken for the best interests of the country as a whole. Hence the duke was able to unify the country though diplomacy rather than through military means. As the superior of Sun Tzu, he also used monetary means rather than warfare to conquer states.

In achieving these objectives Master Guan repeatedly stressed that a ruler's actions must be as genuine in outward graciousness as they are in inner virtuous intent. In other words, national political strategies must be founded on ethics, virtue and morality and a leader must let courtesy and self-control rule his actions. Guan Zhong advised Duke Huan, "Summon those who do not trust you with courtesy and cherish the remote with virtuous conduct. As long as your virtuous conduct and courtesy for others never falters there will be no one who does not respect you." Strategic plans that require the cooperative help of many others will not succeed if their leader does not embody virtue, employ benevolent means as his natural tendency, and treat others with such respect. When you behavior is consistently corrupt or devious then others will not trust you or follow you.

Master Guan was therefore more insistent on virtue than the Chinese strategist Chen Ping who helped Emperor Gaozu of the Han dynasty unify the nation. Chen Ping helped the emperor engage in various tactics of political trickery and thus predicted that his own descendants would not fare well since "secret schemes are prohibited by the Tao school." This was like the modern Chinese saying that those who own casinos will not have good families because of all the bad karma they thereby initiate by running a gambling establishment. Master Guan, on the other hand, fully recognized that if you employ the "yin" factors of deviousness in your affairs then you will bring about your own destruction.

Leaders who simply act according to their own selfish inclinations—who aim to increase their personal power through military force and various manipulative schemes that benefit no one but themselves—can never gain true public backing. A leader must develop a reputation for having a strong moral compass and living according to strong moral principles. Whether from the practical or idealistic perspective, Master Guan noted that without righteousness on one's side and trust and confidence from the public that shows faith in their leadership, political leaders could never earn the solid support of a populace. Thus they could never become true "Protectors of the Realm." True Lord Protectors built their foundations on virtue and public allegiance.

Wise leaders throughout history have all recognized this simple lesson that they need the willing support of the people, and know that military might and authoritarian rule can never secure their genuine obedience.

People typically repose their faith, trust, confidence and allegiance only in individuals of high effectiveness and moral authority. One cannot rely on military power to guarantee peace or lasting conquest because a population must submit to any new rulers, which they will only do if they are seen as trying to benefit the people. Military campaigns can indeed succeed in subjugating a nation, but they are an inadequate recipe for establishing a secure governing relationship within the nation. You must get the people to give you their allegiance naturally.

All people naturally recognize that "Might is not right!" and an indignant populace will always defy the policies of a tyrant, waiting for the chance to break free of his hold. When a nation no longer fears its ruler's oppression then the exercise of authority is sure to meet with defiance and revolt. As Master Guan pointed out, rather than using force it is therefore far better to master the usage of human psychology and political (and public relations) strategies in order to win control of a realm. There is even a Chinese military maxim that captures this very thought: "It is better to win hearts than it is to win cities; it is better to battle with hearts than it is to battle with weapons."

Thus in the *Guanzi* we have many lessons for becoming the supreme leader of a country or group of confederate states—lessons quite relevant to the United States, Canada, Russia, European Community or BRICS union of today. Foremost in this advice is a leader's need to achieve popular support among the people so that his policies are accepted and his government gains firm foundations.

Those who aspire to leadership must first conquer men's hearts and minds, and to do this they must exhibit virtue, justice and fairness in their actions. They must demonstrate a bedrock of moral principles so that they are publicly seen as being virtuous, good people. The goal is to win the trust rather than the anger, contempt and indignation of the people, which is how many Trump supporters feel about president Biden who they believe stole his election. Whether you are the legitimate heir to a throne like a Prince Charles or are trying to become an *elected* official by winning public support, the best pathway for coming into power is through the display of inherent virtue that genuinely wins the people's trust.

This is why Lee Kuan Yew of Singapore would ceaselessly campaign about his party's efforts to help the people, which was to help reduce voter support for Singapore's communist party and its promises. You can't fake these things—they must be real or people will eventually see through them. On the other hand, leaders who claim allegiance to high principles and lofty motives, but who act hypocritically, will gradually lose their adherents and eventually be condemned by history for their disingenuous motives. Propaganda campaigns can only last so long before people become inured to them.

15

To gain influence, it is imperative for an aspiring leader to promote a simple theme which everyone can understand, which is an ability aptly illustrated by the example of Ronald Reagan. For instance, Ronald Reagan would consistently use the phrase "a shining city on a hill" to promote his vision of America and develop national confidence in his international policies that involved the expansion of freedom through the containment of communism.

To achieve popular support you must publicly communicate a grand vision like this in tune with the nation's will that is also aligned with the core DNA of the essence of its culture. You must give the public an enticing vision they can appreciate and with which they can easily identify.[4] You must communicate a message which everyone can understand and then develop a public consensus to follow that vision so that everyone will follow you.

Lee Kuan Yew said that you have to paint a version of the future to your people, convince them it is worth supporting, and then galvanize them to help in its implementation. President Dwight Eisenhower also said that you had to give the people something in which to believe and then get the people to really want to do that.

The highest path is to embody your message in your very being so that your words match with your actions and demonstrate your sincerity. The ideal is to set yourself to a higher purpose and to represent that noble vision in yourself because then you will naturally attract others to rally around you. You want to represent an ideal that is a beacon and catalyst that will evolve the collective entity of the nation to improve its stature in the grand scheme of things.

---

[4] This vision must communicate a central message or theme that is a clear, persuasive story. Furthermore, it's best when all the individual components used to explain this central vision interconnect, interpenetrate and build upon one another. A leader's central vision should promise a more satisfying individual and group identity. The best vision is one that is the leader's central mission that also embodies the contours of the leader's own life. Because it's his very life, this type of story can be promulgated for a very long time. The main message is the most effective if it doesn't require a lot of explanation or education, and is more readily accepted if it can be built upon stories that are already well-known. A leader must not only develop this main message nuanced in different ways, but he must also learn how to deliver this message to his audience. Since a leader's goal is to bring about change in some large and heterogeneous group, it's obvious that a well-framed message is necessary for eliciting support. But the timing behind the delivery of the message is even more important, for a vision must match with the times. The most mediocre message, if it's delivered to a ready audience, will be met with a stupendous response whereas the most eloquent message will fall stillborn if the audience is not ready to hear it. The message is crucial, and its timing is vital.

The highest objective should entail improving the standards of living for the majority of people by helping the nation materially advance, inculcating within society the social qualities that will be useful in the building up of society, producing order and justice in the relationships between person-person and person-state (while preventing monopolies, the powerful or elites from oppressing or exploiting the common people), helping to release a wide variety of creative interests that aid in the culture's positive advancement, and enabling the people to enjoy the maximum of personal freedoms compatible with the freedom of others in society.

It is all about improving the general welfare of the nation rather than the welfare of special interests. One should strive to produce the maximum happiness and well-being for the maximum number of people. One should hope to advance the spirit of the age to a new point of excellence. The ideal is to accelerate in a natural way – by using our knowledge, wisdom and intelligence and working with the natural drives of humankind – the cultural and material prosperity of the people by bringing them to their highest state and fullest expression of potential.

A true leader should want to act under the influence of an ideal, the purity of which can guide the nation, so that the people will embark on a course of development that progressively realizes that noble ideal and brings out the country's magnificence. The objective is to set in motion a heroic manifestation of service to the highest principles that makes the people great in order to make the nation capable of greatness.

If your personal identity merges with the essence of the message that you are trying to convey, the public will then have confidence in your sincerity. You have to live your talk, and this is how you'll win the confidence of the public. Those without the support of the common man will always stand on shaky ground, and when a leader's position is insecure others will chatter about their personal failings and moral deficiencies.

The *Guanzi* thus details many strategies for gaining power and influence, applicable regardless of whether one wishes to become the supreme leader of a group, a nation or a confederation of states. The lessons are aimed at politics and rulership, but they can just as easily apply to the world of business today where corporations are like little states in themselves. These are lessons on leadership and strategy that can be applied almost anywhere.

If you decide to study the *Guanzi* and master its lessons, as Japanese strategists of realpolitik have often done, you must not overlook the moral dimensions of the work which are often neglected by those who just want to gain power and control. In truth, no one can really succeed nor long prosper with grand schemes unless their plans are soundly founded upon virtuous intent. If policies are not developed to help human beings rather than just take from them then they will just disintegrate over time.

Yin is always replaced by yang so in the long run evil cannot prosper, and prosperity will self-destruct as well when the virtues that built it up become forgotten. The stronger that your plans are based upon the realities of human nature and man's desire to improve himself both economically and culturally the stronger will be your success. Marxism and Communism, for instance, are grand failures along these lines and always will be failures no matter in what form they appear in the future.

Master Guan is not to be considered a heartless, scheming Chinese Machiavelli because he loved the people and always emphasized that a leader treat them with concern. Master Guan felt that leaders should guide themselves privately and publicly by just and moral principles for these are the foundational principles behind any well-run organization. Adopting some false ethical stance, taken for expediency's sake, would just not do. Everything, he stressed, should always revolve around virtue.

As most recognize, there is no amount of propaganda nor public relations which can mask a lack of virtuous intent; it is impossible to adequately camouflage or disguise selfish actions for long when they lack a moral basis. Those who act only in terms of their own self-interest, lacking virtuous motives while fueled by a greed for money or lust for power, will always be found out in the end. It is simply a matter of time before this inevitably happens. The final reckoning and a subsequent loss of support will come sooner or later.

One can always rely on the public's common sense and collective wisdom to spot falsity and identify who isn't sincere about a better future for the nation. Great leaders—leaders who are great—are the ones who always operate with a mind devoted to enriching the people, protecting their interests, promoting the public good, and saving the nation. This motivation is the pillar around which they evoke the support of the people.

Even though Master Guan emphasized benevolent government and dissuaded from the use of force unless it was necessary, this did not mean that he was a pacifist when it came to the topic of war. He used economic warfare to conquer surrounding states and advised his duke to attack the neighboring state of Xing that was under attack from nomads. He was not totally averse to military undertakings but simply insisted that other methods were more effective, less risky and less costly. Military campaigns are fraught with dangers, and expensive in terms of lives and economics. They are the realm of chance—unpredictable beasts to the very end. No ruler should wish to risk his kingdom on the battlefield so Master Guan emphasized a different way which was conquest through economics.

In Guan Zhong's view, military might was an instrument of policy to be used like any other, but strength had to be subordinate to the idea of right. Because he wanted his lord to become the most respected and powerful in China he warned that military activities should never be countenanced without righteousness, and naturally they were to be avoided if the chance of losing was greater than the chance of winning.

Since nothing can cause greater loss than war, where a single day's fighting can destroy an entire generation's worth of labor, Guan Zhong adopted the realistic policy of being careful about entering into any military engagements. There are always a variety of alternatives to an end, including a multitude of ways to win without fighting. Guan Zhong's guiding principle was that his state should "be strong for war if needful, otherwise practice benevolence."

Guan Zhong was not an idealistic dreamer who felt a nation could survive if it left itself unprotected. He emphasized that his state of Qi should build up its economic strength and achieve greater internal stability because these two characteristics were often the best deterrents to invasion. A wise nation, he felt, must always be prepared for war, but like all other great strategists he insisted that the best strategy was to win without fighting,[5] or else to win in the quickest and least costly manner as possible.

To become a supreme nation or supreme leader it has never been prudent to put blind faith in military force, but rather, it is always much wiser to rely on insightful strategy. With good strategy and good timing you can achieve much more with less.

Sun Tzu pointed out that those exhibiting the very highest skills of warfare are those who never fight with anyone. Guan Zhong therefore developed a number of strategies for making his own state strong so that it could subdue opponent states through economic and diplomatic means rather than through warfare.

Today it is popular to study the writings of Sun Tzu, but many of Sun Tzu's writings simply paraphrase the *Guanzi*. While Sun Tzu and Guan Tzu came from the same state and commented on similar topics the principles and strategies within the *Guanzi* far surpass Sun Tzu's teachings in terms of the deep understanding required of leadership and geopolitical affairs.

Despite the mastery of warfare and military maneuver shown by both of these masters, it is important to once again emphasize Guan Tzu's insistence on Sun Tzu's most important maxim: "To win without fighting is best of all." One should not misconstrue this small volume as a blueprint for militaristic or imperialistic designs.

In addition to rivaling Sun Tzu in his mastery of strategy, it must be mentioned that Guan Tzu's civil and administrative policies were on par with those of Confucius who appeared over 150 years later. In fact, much

---

[5] Confucius might have said exactly the same thing, for in *Analects* XII.13 we find him commenting, "I could try a civil suit as well as any other man, but it's even better to bring it about that the parties don't resort to litigation at all." This is the same principle applied to civil affairs.

of Confucius' personal administration mirrored the substance of Master Guan's teachings. He also acknowledged Master Guan's great merit in that he had enabled Duke Huan to assemble the feudal lords nine times without need of force, and it was Guan Zhong's administration which had saved the Chinese Empire from collapse. In effect, Guan Zhong's reforms had prevented the Tartar barbarians from conquering China thus preserving the Chinese civilization from dissolution. Hence the Chinese are still enjoying the benefits of his actions today.

These accomplishments bear witness to Guan Zhong's political, administrative and military abilities but say nothing about his contributions in the "soft" field of social philosophy. However, anyone who bothers to investigate the *Guanzi* carefully will find many sections which rival Mencius. It is also packed full of Taoist thoughts pre-dating Lao Tzu and Chuang Tzu. Some of these ideas are from latter contributing authors so we cannot attribute all of them to Guan Zhong.

Turning to economic matters, Guan Zhong originated a number of novel economic policies which gradually made Qi into the richest and most stable state in China. He reformed the agricultural taxation system by replacing the traditional land rental with taxes levied in accordance with the size and quality of the land under cultivation. This acted to stimulate Qi's agricultural and silk production. It also freed the peasants from the crush of a burdensome and ineffective tax system which did little to help fill the state's coffers.

Guan Zhong also organized the citizenry into special work groups by classifying them as either traders, workers, farmers or scholars. By relying primarily on the economically unproductive as soldiers in times of war, while shielding the more productive farmers and commercial classes, this new strategy enabled the state of Qi to stay productive during both times of peace and warfare. He encouraged agricultural production, reformed the tax system to be based on productivity, and discouraged private extravagance while encouraging good behavior by effecting a system of rewards and punishments. Guan Zhong nurtured an environment conducive to the work ethic thus enabling the state of Qi to lay strong economic foundations and accumulate the resources necessary for military expenditure.

Like the economist Lord Keynes, Guan Zhong sanctioned the principle of government economic intervention, namely fiscal policy, to strengthen the national economy. He specifically encouraged public works in times of recession. He advocated the government management of key industries such as salt, iron and timber production and taxation of those industries in order to avoid excessive direct taxation on the populace.

Master Guan was originator of the policy of "balancing the light and heavy," which was basically the regulation of the money supply to control inflation, and he advocated the careful manipulation of prices and money by the state. This made him one of the world's first Monetarists, who clearly recognized that the circulation of money and the variability of prices influenced the supply and demand of commodities as well as employment and immigration. Guan Zhong also used his understanding of monetary policy (the light and heavy) to strategically inflict damage on opponent states without resorting to open warfare. Guan Zhong's successful combination of Keynesian and Monetarist policies still offers useful lessons for policy makers today.

In short, in serving as the chancellor (chief or prime minister) for Duke Huan he instituted a series of economic, administrative, social, diplomatic and military measures which fostered the welfare to the people, and which gradually turned Qi into the foremost state of China.

Relying on his advice, Duke Huan was able to form a strong confederation of the previously disunited and uncooperative feudal states. This confederation professed a renewed loyalty to the emperor and repulsed the invading barbarians of the North. In restoring the dignity of the emperor and the imperial privileges, forcing the leading vassal states to fulfill their liege responsibilities and putting a stop to the local wars of ambition, Master Guan was responsible for resuscitating the moral basis of the imperial system and for reunifying the Chinese nation.

Guan Zhong's policies and actions during this period made Duke Huan the first "Lord Protector" of China, but the sage and master strategist Guan Tzu should be considered the actual savior of the nation.

What Guan Zhong has left to us in the *Guanzi* is therefore a masterpiece on economics, military affairs and overall statecraft, an ancient work of geopolitics with few equals. Though Duke Huan, as ruler of the state of Qi, always retained his right to question and alter Guan Zhong's policies and strategies as he saw fit, Guan Zhong's success across the entire spectrum of government affairs lead the duke to place great trust in most all of his recommendations.

It is ironic and unfortunate that someone with such great abilities was not used by his previous superiors, but such are the twists of fate that the great sometimes go unnoticed until it is their time. Such was the case of Jiang Ziya, the famous strategist and author of the *Six Secret Teachings* who helped King Wen and King Wu overthrow the Shang dynasty in ancient China.

As to the effectiveness of Guan Zhong's policies, Qi's ascension to the position of preeminent state of China in itself speaks of his wisdom and statecraft. That's one reason that his policies deserve a careful examination today, so that all the world's nations might benefit by them.

CHAPTER 3
POLITICAL LEADERSHIP AND SUPREMACY

LESSON 1: What Qualities Distinguish the Great Leaders?

There are certain conditions which distinguish the lord protectors and kings. They model themselves on the workings of Heaven and Earth. They educate the people and transform society. They establish laws and regulations for the nation and a hierarchy among the feudal lords. They treat everyone within the four seas as guests and always act at the proper time. By acting in this fashion they are able to restore law and order to the whole world.

In order to lead the world you must love the people as if they were your own children. You must desire to create a brighter future for them and pathways to happier lives by working to open up opportunities and ending their sufferings and afflictions. To become a great political leader you must be genuinely concerned with the welfare of others. You must be gifted with compassion and a commitment to nourish all the people of the earth. The universal goal of benevolence entails universal impartiality and universal compassion.

Confucius said that the first thing a ruler should do for a people is to improve their circumstances—to enrich them and bring them out of poverty. You can nourish them with ethics and culture only after their bellies are full. Most leaders are elected today because they promise to improve the economy and lead the country to greater prosperity. Prosperity is the basic issue of a country because it is through prosperity as a foundation that a domain can be made orderly.

When we talk about "helping the people" and "contributing to society," the first thought everyone has is to make the people more wealthy, such as by developing the economy so that businesses thrive, farmers thrive and the people can become enriched because they do not lose too much through taxes and regulations. Maybe we can provide the people with more material wealth or the means for wealth, but whether this will make them happier is another question entirely. Even so, Confucius' point is that it is harder to instill virtue and peace in a nation when the people are idle and hungry. You must first fill their bellies before you can lecture people on ethics.

When society is impoverished and hungry it is hard to establish the rule of law because people will do whatever they must to survive. Without the social order of a settled society the operation of law is impossible. Therefore one of your primary goals should be to establish conditions that improve the standard of living for the majority of people, and then inculcate in society those social qualities that will be useful in the building up of the state.

This touches upon educational matters as well as the virtues stressed by the culture and religion. The purpose of the national educational system should be (1) to train people in *particular skills with which they can make a living* so that they are able to live in the world in an independent, self-directed, self-reliant way, (2) to produce *good citizens* who care about their country and are willing to strive for the public good through civic contributions, (3) to produce *good, virtuous people* with ethical standards who can live peacefully with others in harmonious, cooperative human relationships.

The TED presentations of statistician Hans Rosling (available on Youtube) can guide you through many basics of laying the foundation for a prosperous society.

Confucius said that the second task of a leader was then to educate and instruct the people, which means teaching the people ethics and virtue in order to establish peace, law and order. First you take care of economics and then you take care of moral education. You need to teach the public the virtues that will enable people to cooperatively live together in peace and harmony, and which will help in the building up of society. The target is to not just to bring people out of economic misery but to elevate the nation's character and social culture. Ultimately you want to release a wide variety of creative interests so that your nation attains a high measure of material and cultural advance.

A political leader must teach the people as if they were a new generation to advance his nation to a new point of excellence. He must work to uplift their sense of identity by bolstering the core values and positive DNA of the culture that leads to magnificence.

A leader must also be willing to promote innovative changes that go against outmoded traditions. He should nurture excellence and encourage the average to improve. He should work to develop men and make them great in order to make the nation great. He should reward those who are good and correct errors in the public body.

However, during this overall process a leader's private actions must be consistent with the values he publicly advocates. He needs to represent in his own person (not just his speech) an ideal whose purity can guide the nation, and which can serve as a beacon that evolves those qualities that will permit the nation to improve its stature in the grand scheme of things.

In order to lead others, it is absolutely important for a leader to develop plans which the people can follow so that the country doesn't stay at the same stage of development for too long. Abundance occurs, for instance, because of a well-designed intention followed by well-designed actions taken over and over again as a plan until the effects of those repeated actions replace the consequences of the bad actions of the past. A leader's job is to derive those plans and then get the people to really want to institute them.

A leader must instruct the people about these policies and how they fit into an overriding vision so that everyone understands their purpose and he gains their willing cooperation in the implementation. The objective is a heroic manifestation of voluntary service to the nation's highest fortunes. The target is that the nation progressively realizes a noble ideal that brings out its magnificence. To accomplish this a leader must promote an ethic of service and sacrifice.

In instituting these policies an organization's governing system should be extremely fair so that all people are given equal consideration, are treated fairly, and no one is looked down upon as an inferior. As I pointed out in *Culture, Country, City, Company, Person, Purpose, Passion, World* as regards cities but which also applies to a country as a whole: "A city aimed at prosperity must present itself as legally and economically fair to all participants and be administered according to a clear and impartial set of laws. ... The impartial rule of law (and fair taxation system) insures that one's assets are not stolen by the government, competitors or elites, who historically tend to oppress weaker parties that cannot protect themselves. Without a dependable legal system a city will not see strong capital formation and economic development. Without the rule of law, few will risk their capital to go there. Minorities should (also) be given equal opportunity to excel, and given access to a level playing field of opportunities and law enforcement."

To illustrate the policy of fairness we can turn to the example of the Roman Empire since it consisted of a large number of vastly different territories. Although the Romans were known as conquerors, which tends to conflict with the idea of being "fair," the Roman administrative system was flexible, tolerant and open. Virgil wrote in the *Aeneid* that the great art of the Roman Empire was, "To govern the people with authority and establish peace under the rule of law." The rule of Roman law is what produced fairness.

Rome was known for its equitable taxation policies and its rational framework of justice over the lands that the Empire controlled. While the Empire was composed of over forty provinces peopled by different tribes and nationalities with entirely different customs, languages, practices and traditions, the Romans respected these separate cultures, gave them real authority, and let them manage their own affairs. Under the short-lived policies of Alexander the Great, on the other hand, the defeated were treated as enemies whereas the Romans treated their provincial subjects as Romans and as contributors rather than outsiders. Hence the local elites in provinces were given important positions in the Roman hierarchy, and most every position in the Empire was open to a suitable candidate irrespective of his origin.

This attitude is to be contrasted with the old policies of the British Empire wherein the British considered themselves a culture superior to the colonies they managed. As a general rule the British would never fill important administrative posts with local citizens. One only need compare the longevity and reach of these two Empires, and you can add a comparison to the Empire of Alexander the Great to the mix, to understand which means of governing constituted the wisest course of action. Alexander wanted people to view him as one of two extremes – kind, caring, heroic and merciful ruler or a cruel, bloodthirsty madman and tyrant – in order to create within his own men and society a psychological need to follow him no matter how they viewed him.

If a nation wishes to lead the world it must learn to appreciate local cultures instead of maintaining that its own cultural, political, and ethical standards are best. Many British administrators in India, as an example, recognized that its culture was older and in many ways superior to the British Anglo-Saxon order.

In short, everyone should be equally respected because the principle of prosperity is to establish a meritocracy regardless of people's backgrounds and to allow social mobility so that people can freely climb the social hierarchy and improve their economic standing. Some large multi-national corporations therefore wisely promote local leaders to manage their foreign operations because they understand those conditions better than the head office. Guan Zhong broke with tradition and steered administrative responsibilities away from aristocratic clans to trained officials, which was one of the reason Qi was able to become the most powerful state of his day due to his governance.

Thus a supreme leader treats others with equal consideration as if they were all important guests. When leaders treat everyone fairly and value their input they will gradually acquire the necessary allegiance and influence to be able to steer their organizations in necessary directions.

Using these principles a supreme leader will then become able to successfully correct what's wrong in the world. When a leader acts for the sake of the world he draws the world to himself, and when he consistently shows his love for the world he subsequently wins the trust of the world. This is why he can correct great matters. When you are loved by people who trust you and put their faith in you then you can accomplish even gargantuan tasks due to receiving willing cooperation from those you are serving.

Several other Chinese classics that may help in this task which I recommend include *T'ai Kung's Six Secret Teachings* (Jiang Ziya), *The Three Strategies of Huang Shigong*, Laozi's *Tao Te Ching (Dao De Jing)*, and the *Thirty-Six Stratagems*.

LESSON 2: The Challenges to Becoming the Supreme Leader

> There are several tasks to accomplish [when a lord protector or king wishes] to become supreme. They reduce the size of large states and correct those states which behave in an unprincipled manner. They cut down the might of those states which are excessively strong and reduce the influence of those states considered important. They bring unity to states racked by rebellion and dispose of rulers who have indulged in violence. They punish the guilty, demote the unworthy holding rank, and become the new leaders in order to sustain the people and establish them again.

To become the supreme head of a nation or leader of the world requires skillful strategy, patience, timing, and effort. Because strategy is a primary concern Master Guan therefore advises several tactics for becoming the eventual leader of a large realm of many states. These tactics involve gaining political power and include cutting the larger states down to a smaller size, curbing the military might of states which are too powerful, and decreasing the stature of states which wield too strong an influence over others.

As in warfare, this is the practice of trimming an enemy's strength by cutting away at it like a butcher who dismembers a carcass. A present day American and British plan to divide Russia into a dozen or so member states with leaders secretly appointed by the West follows this game plan. The idea is to cut it apart to weaken it, control each section so that they continually squabble with one another, and plunder its fabulous resources.

The reason that trimming strategies work is because they rely on the inherent principle of weakening by dividing or diminishing. By decreasing the strength of any elements which are someone's advantages you will then be able to more effectively use your powers to deal with them more easily.

Those who aspire toward a supreme leadership position commonly admonish and discipline rulers who exercise absolute power in a deviated fashion. They offer themselves as a refuge to the oppressed so that they also serve as protectors to the fallen. When they can restore equilibrium to political situations by replacing the rule of tyrants with benevolent administration then there can be peace in a realm rather than disorder. A "realm" may represent a country, business or other type of organization with many members.

Political leaders aspiring to worldwide influence should replace their own personal desires, and likes or dislikes, with the objective of helping the whole world. They must rescue states facing extinction whose survival is in jeopardy. However, supreme leaders should never intervene in peaceful countries with an intention to harm the reigning regimes as the U.K. did when its intelligence apparatus tried to help defeat President Trump's election and remove him from office. They should only intervene in suffering nations with the intent to save people who are on the brink of destruction. This is why they assume control over states misruled by tyrants and dispose of those despots.[6] A Chinese maxim aptly says, "Launch an expedition to punish the sinner and save the people."

A supreme leader's military interventions should be motivated by benevolence rather than by the desire for conquest or geopolitical gain. In such cases, Master Guan also feels that a responsible leader should protect and govern the fallen until they are once again prosperous and able to raise their own leaders. If one is a hegemon, on the other hand, he will continue to rule over the people indefinitely because the intervention was never motivated by benevolence. Despite pretty words that a nation is "intervening to save people" you must always be suspicious of the true intentions behind interventions. In today's world, most such claims are fake public relations efforts used to hide ulterior motives. There is always a rationale provided to the public that sound great and a set of *real geopolitical objectives* behind the facade.

Strategic interest, or the call of foreign alliances, is the most common

---

[6] The task of toppling tyrants to alleviate the people's suffering is even sanctioned in Buddhism, which is known for its pacifist policies. The Buddhist sage Asanga said that a savior might even kill a king if this is the only way one can stop him from committing atrocities. As a general rule, by the way, one should never mix politics with religion. History easily shows that religious dogma has always been a poor guide for national politics and that those countries which mix religion and politics typically become internally weak.

reason for becoming involved in a foreign intervention so a leader looking for political power cannot rest in isolationism. It is far too dangerous and does not give him access to greater power. To become a supreme hegemon a ruler must go out of his way to interact with foreign countries on a regular basis.

For instance, if he truly loves the world's people (is worried about their welfare) and wants them all to prosper a leader will offer help to those who have been harmed by catastrophe such as war, famine, flood and so forth. Out of benevolence and good relations he will assist them to rebuild again even if there is no strategic value to these actions. Eventually, states suffering temporary difficulties will be able to stand on their feet again and will remember the assistance they had been rendered. Countries have long memories when they help they have been rendered was significant. By disseminating charity where and when necessary a leader's greater goals can be accomplished with more ease.

Only if you love people, rather than power, should you be motivated to try and become an influential world leader. Furthermore, despite compassion and the desire to help you must learn great restraint so that wisdom and strategy guide your actions, and you must also learn skillfulness. In Chinese culture you would study several brilliant figures to learn strategy, wisdom and skillfulness including Jiang Ziya, Zhang Liang, Zhuge Liang, and Guan Zhong.

Wisdom, rather than emotions, must be your guide when you are trying to achieve grand objectives. You must embrace a fact-based worldview, be aware of your cognitive biases, remember that trends do not continue in a straight line forever, take a wider global perspective, question authority and always look for opportune timing. Furthermore, not all the problems in the world can be solved, nor should even be addressed. In some cases any type of interventionist "cure" will be worse than the original problem. Doctors commonly perform triage to determine the order of priority for treating medical emergencies and some national or world problems do not have high priority for your time and resources.

Good leaders establish priorities for their activities. Since they are aware that some objectives constitute futile undertakings, some require too much effort, and others will naturally resolve over time they carefully refrain from over-extending their efforts and commitments.

LESSON 3: The Difference Between the Superior and Inferior Leader

The ruler who merely enriches his own state is qualified to become a lord protector, but one who is righteous is qualified to become a king. A true king should possess great wisdom. He does not attack states that uphold the same virtues as his own nor establish his rule over those that are alike in adhering to the Way. However, among those who are contending for dominance of the realm it is common practice to use force to overthrow tyrants.

There is a great difference between the leader of a nation and one of its minor politicians just as the president of a corporation greatly differs from its divisional managers in terms of status and abilities. The great address the larger while the lesser address the smaller.

A minor leader is typically preoccupied with concerns subordinate to group interests as a whole. He pursues strategies which focus on smaller objectives and addresses the limited scope rather than the greater. The Congressman or divisional manager both fall into this category because their efforts are mainly targeted at parochial concerns and, by definition, individuals who simply address minority interests are not majority leaders.

Each group within an organization or country typically wants its own in-group interests to dominate over everyone else. To become a powerful political leader you must learn how to balance all those minority interests with the interests of the whole so that you can lead the organization to greater prosperity and power.

To become a supreme leader a man must expand his perspectives to address the larger issues of prosperity, growth, strategic risk and survival. His viewpoints must envelop the whole and in order to triumph over competitors his morality must also be higher than that of other contenders for power as should be his wisdom, reliability and skillfulness or effectiveness.

Unfortunately, all of this places a great burden on the leader seeking political power because the larger the population the greater will be the challenges he faces in crafting a story with audience appeal. This message he needs to transmit must tap the nation's interest and be a simple one embodied in his personality and behavior or he will be seen as a hypocrite. *Culture, Country, City, Company, Person, Purpose, Passion, World* states that the grand strategy he should pursue should embody the following characteristics:

> A vision rooted in human nature and the deep essence of the country (culture, group, empire) that is so noble and inspiring that it not only attracts the uncommitted and lethargic, but also undermines the opposition of pessimists, cynics, oppositionists and other adversarial naysayers. This grand notion of shared purpose should be so attractive and compelling that the entire

country (culture, group, empire, community) becomes united and committed under its ideal.

The mission promise of magnificence should enervate both the inner and outer life, and entail an intergenerational contract of social motivity. It should act as a beacon, catalyst or guide around which to evolve in the nation those qualities that permit the collective whole to improve its stature in the grand scheme of things. It should insure that the community does not stay too long at one stage of development but decidedly progresses to advance forward.

Allowing people to connect themselves to a destiny story greater than themselves that resonates with glory, it should magnify the spirit of adherents who share in its participation, yet allow everyone to freely retain their own individual distinctive characteristics. It permits and empowers every person to enhance and perfect their own unique skills, interests and contributions within the larger narrative of the cohesive guiding vision.[7]

For a nation to become a superpower it must adopt sound policies—forged of wide vision and wisdom—that will bring wealth and prosperity to the entire nation because it is primarily rich countries that become superpowers, and no one stays supreme forever. The leader who has the wisdom to raise up and enrich an entire nation rather than just some tiny internal group (and the executive who can make his entire company flourish rather than just some division) is qualified for the leadership helm. He applies his efforts to insure his organization's security and benefit the interests of the whole.

However, just as you must possess greater vision and perform more effective actions than your opponents in any competition, those aspiring to the leadership helm need incredible resources, power and the wisdom of a sage to rank supreme. They need the ability to carry out their vision without outwardly appearing to be despots, but this can only come from willing public support that is won because they are seen as virtuous, benevolent individuals interested in the people's welfare.

It would be nice to think that virtue is sufficient in itself for an individual to rise in life. However, those who become successful leaders usually know expert means of motivating others through persuasion, accumulate lots of political capital, and who actively maneuver to obtain power and influence. Like successful movie directors, parents, managers

---

[7] *Culture, Country, City, Company, Person, Purpose, Passion, World*, Bill Bodri, (Top Shape Publishing, Reno: Nevada, 2018), p. 26.

and even dictators, great leaders are superb motivators who can influence people toward much greater ends and gain their support for their own projects.

In the struggle for supremacy there will be many contenders jockeying for power. Wealth, status, privilege, dominance, recognition, sex and power are the things that men most like and many will pursue power to satisfy these other wants. Master Guan insists that those competing for positions of power should not try to overlord contenders who share the same values because this will alienate your peers who are by nature natural allies. If another country is following a similar political system as yours then you should not try to overlord it by using military force but find gentler ways to influence it as the U.K. does with the United States. It would be ludicrous if France were to launch a war against Germany, Spain or the United Kingdom because all these countries are democratic-capitalist peers who can thus be said to share the same principles and "virtues."

On the other hand, "sharing the same virtues" does not mean holding exactly congruent views or political systems. It does not suggest that countries holding the same basic values will be reliable or consistent allies either because each wishes to chart its own way and become dominant over the others. Each is interested in taking trade away from competing nations because trade is the source of every nation's wealth and prosperity.

Countries become allies because of common interests, but in a world where countries are trying to outwit one another for international trade and investment flows one's allegiance can change according to one's self-interest. It makes sense to stay on friendly terms with your competitors and *not alienate your allies* who typically support your common interests.

For instance, the Peloponnesian War was effectively a rising up of city-states that were once the allies of Athens and who joined Sparta against Athens because of its arrogance. In the Italic War (91-88 BC) there was a revolt of the central and southern Italian allies of Rome that had fought side by side with it in several wars but were denied Roman citizenship and the privileges it conferred. One must not alienate your allies but stay on good terms with them.

While one should refrain from attacking other parties sharing "the same virtues" the leader who aspires to have supreme influence in the world will not shy away from attacking the aberrant. In the political realm these are tyrants and despots who use violence to harm their populations or others, hence they imperil their own or other states. The United States attacked Colonel Noriega of Panama, Muammar Gaddafi of Libya, and Saddam Hussein of Iraq after first branding them as tyrants and despots.

Overlording the deviant has always been a common practice among those wishing to become world leaders, and who righteously employ force to end a despot's misuse of force while displaying such righteousness to the public. They will publicly use force to correct what everyone knows is wrong and make a big display out of protecting others.[8] Therefore when it is beneficial a world protector shows no hesitancy in taking over other countries politically unsettled due to the misrule of a tyrant, despot or group of villains. Then he tries to nurse it back to health.

Once a leader intervenes to save another nation he must avoid wavering in this commitment. Caution is in order before doing such things because he must recognize from the outset that intervention might be a costly long-term proposition just as was the case of America's participation in the Vietnam war.[9] A leader who practices international intervention and conquest must be prepared to remain with his ward for decades if necessary, perhaps even assuming control over its affairs in order to nurse it back to health and establish it again. Otherwise, if you continually initiate various foreign interventions with hit and run tactics without trying to stabilize those nations then your reputation will tarnish and over time nations will come to mistrust you and then abandon you as many nations are now doing with the United States dollar.

For one thing, a great frequency of military forays will cast doubt upon a leader's true intentions. Propaganda can hide one's intentions only so much before everyone figures out what is really going on. If you are trigger happy and commonly attack others it will gradually lead to even allies questioning whether they can continue to trust you. Hit and run interventions have also often been shown to inflict even greater damage to a situation than whatever existed before their commencement, such as the toppling of Libya's Muammar Gaddafi that led to the reinstitution of slave markets in Libya and massive unwanted African immigration into Europe.

[8] It is safe to say that many African nations, which commonly suffer misrule and strife, must take this warning to heart because many foreign nations lust after their resources. Foreign powers will use any deviances (such as rebellions, insurgents or terrorists) to justify foreign interventions in African nations as camouflage to gain control over their resources. Over the future we will especially see many attempts to gain control over gold mining operations in the weaker nations of the world, especially in Africa, using various forms of cover.

[9] Even before this a nation (or any institution) must make sure its commitments do not outpace the size of its resources, meaning that resources should be adequate enough to match commitments. If a nation overextends itself in multiple obligations and becomes an untrustworthy alliance partner its allies will rethink their agreements. This has happened in the relationship between the U.S. and Saudi Arabia. From the start, the number and type of national commitments must be given careful consideration because they must be able to survive a perfect storm of multiple bad conditions. The simultaneous bankruptcy of many U.S. Savings and Loans in the Savings and Loan Crisis of the 80s and 90s, teaches us that in times of difficulty everything seems to go wrong together.

The consequences of intervention should be considered ahead of time and in this case they were a mistake, but foreign nations lusted after his oil wealth, control over his central bank, and wanted to crush his push for an African gold-backed currency that was independent of the dollar.

Business executives who wish to dominate their industries would receive blessings from Master Guan for taking over companies on the verge of bankruptcy and managing them back to health. By doing so they would become heroes who not only saved working families by preserving jobs but would acquire assets and skills cheaply while strengthening their own competitive position. Such a course of action has both humanitarian and strategic benefits.

LESSON 4: The Tao of Accomplishment

> Ruling over the people has its principles (tao), and becoming a lord protector or king has its timing. To have your own state in perfect order while the neighboring states are in disorder is a major opportunity for becoming a lord protector or king.

There are definite principles to follow for becoming a leader and for being a good leader. One of those principles is to always be sure to act at the proper time. You must act when action is necessary *before it is too late*, and must seize opportunities before they disappear. Timing is the key to achieving more with less.

Those who focus on the grand scale therefore study the timing involved in Strauss-Howe generational cycles, Kondratieff waves, business cycles, John Glubb's stages of empires, Ibn Khaldun's theory of social development, Wheeler's long-term weather-government cycles, Elliott waves and socionomic patterns (*Socionomics* by Robert Prechter), and even mundane astrological timing. The famous number "666" within the Book of Revelations is actually a reference to long-term planetary cycles that rule the affairs of men and a supreme leader strives to understand such things (cyclical change in long-term history) through study of works such as historian Richard Tarnas' *Cosmos and Psyche* and Bill Meridian's *Planetary Economic Forecasting*. The final digit 6 within 666 represents the human 24-hour clock, the 60 represents the 60-year Saturn-Jupiter cycle that rules human affairs (as stressed within the Chinese and Indian cultures), and 600 represents the 3600-year period of the conjunction of Uranus and Neptune.

Incidentally, according to Andre Barbault a Jupiter-Saturn conjunction in 2080-81 passes over a Uranus-Neptune opposition that will be a kind of planetary threat intensification of 1914 and 1942. A Barbault graph indicates that the time period is more likely to be around 2082 whereas a Ganeau graph suggests 2080. Both graphs, which clearly identify prior wars, also suggest a significant war starting in 2031-33 and 2060-61 (see lunarium.co.uk).

A successful lord protector or king always uses strategic timing to achieve his goals because opportune timing allows you to achieve more with less effort. An executive who wishes his proposals to gain acceptance must also master timing for there is nothing more powerful than waiting until the time is right to receive opposition-free approval for one's plans.

Another important principle to put into practice, when you have already become a leader, is to establish a formal means of managing your organization. To govern a country or manage an organization you definitely need an effective administrative structure without a lot of layers otherwise you will not be able to act quickly in response to events. The less layers you have the quicker will be the lines of communication and your ability to act.

Also, when a nation lacks an impartial system of rules (fairness involves opportunities and procedural justice) for regulating its affairs then there is bound to be disorder within the state. When the means for ruling the people lacks any sort of organization there will always be chaos in internal affairs. To establish peace in a large nation you need an underlying system of law and order that everyone can understand and follow. No matter whether your country is authoritarian, republican, socialist, democratic, capitalist or communist, it still requires some form of law and order and steady government administration.

History shows that the greatest success and stability has always come to nations that maintained the social order, educated their people, maintained peace with their neighbors, accorded property rights to individuals, allowed them free mobility, allowed a free market, and governed the people according to an established rule of law. This allows individuals to plan their affairs and take calculated economic risks because they can then know what treatment to expect from the state. On the other hand, when an environment lacks clarity because people don't know what to expect this tends to stifle entrepreneurial risk-taking and capital investment. A saying runs, "even fish have trouble navigating when the waters are extremely muddy."

While a government needs an efficient organizational structure for administering its internal affairs it must make use of strategic actions and timely opportunities to promote internal policies and its global standing. For instance, a nation that wishes to become greater will use its various forms of influence to overtake other states politically unsettled. While relying upon the law to ensure order within itself it will take advantage of the difficulties suffered by neighboring states to advance its own positions.

When the enemy of one's state suffers some casualty the wise political leader of a realm sees this as an opportunity for national advantage just as a corporation watching a competitor make some strategic blunder will take advantage of the mistake. Business firms will commonly use competitor mistakes to strengthen their own positions. Some companies will try to expand their market share in such situations while others might try to destroy a beleaguered colleague and eliminate that competition forever. The competitive principle is that a rival's difficulties can become your advantage if you are smart enough to use them. When they are doing something stupid then just leave them alone so that they dig for themselves a bigger hole.

In Taoist terminology, Master Guan therefore advises a leader to take advantage of a neighbor's yin tide to increase your own yang fortunes.[10] In other words, a neighbor's instability is your own opportunity for gain; a neighbor's misfortune can be your good fortune if you seize the chance.

To become the most powerful leader of all you absolutely must seize such chances because opportunities rarely repeat themselves. In geopolitics you must not fail to make use of any such chances that come your way because your competition tends to be smart enough to minimize its mistakes. Such opportunities offer a lucrative situation where you can become an even greater power without any great effort or risk to your own position. The Tao of accomplishment is achieved when one makes use of such opportunities.

LESSON 5: Chaos Means Opportunity

> The survival or ruin of a state is greatly influenced by the conditions of its neighbors. If a neighboring country experiences difficulties then one might reap some benefits or suffer losses on that account. When the entire world experiences instability and disorder this can be used to a wise king's advantage. Should his own state be in danger, however, he would be aware of this through his wisdom. The ancient rulers became kings by taking advantage of the misrule and injudicious actions of their neighbors. The enemies of these neighboring states were also pleased by such actions.

[10] Yin and yang refer to a duality of opposite but complementary characteristics that appear in nature and alternate in turn. A large set of characteristics are attached to the concepts of yin and yang where yang typically refers to "positive" and active characteristics while yin refers to "negative" or receptive qualities. In this particular example a superior strategist takes advantage of another's disorder (yin) to increase his own benefits (yang). In other words, he takes advantage of someone's bad fortune to increase his own good fortune.

In ancient Chinese philosophy, opportunities arise from crisis for those who are prepared and willing to act. Such is the advice heard time and again in the investment field where the greatest gains often come from swift actions taken during times of crisis when, as Baron Rothschild once said, "there is blood in the streets." Rothschild made a fortune buying during the panic that followed the Battle of Waterloo against Napoleon when most investors lost their heads in panic.

One way in which leaders with power distinguish themselves from ordinary men is by making use of (and gaining through) troublesome situations, including an opponent's difficulties, because their troubles often equate with your opportunity.

You can discover a silver lining in a raincloud if you know how to turn an unfortunate circumstance to your favor, and to become a great leader you must learn how to do this. Instability can mean opportunity for the smart and quick, the courageous and the prepared who are willing to act when things seem dangerous.

If a leader has the tendency to hesitate during opportune moments, however, he will never be able to achieve any grand objectives just as Guan Zhong taught Duke Huan when they first met. We must remember that it is the investor who buys when "blood is running in the streets" who has the best chances of becoming rich because that is when prices are suppressed due to emotional overreactions. He knows that his window of opportunity is rare and may last for only a brief period so at those times he seeks investments in companies with lasting value because prices will rebound after emotions stabilize and deviances revert to the mean.

An entire field of investment, called contrarian investment, is built upon this notion of going against prevailing market trends to sell when others are buying and buy when others are selling. Warren Buffett is a famous contrarian investor and my book *Super Investing* contains various models used by Benjamin Graham and others to employ this investment philosophy. Many of these models have worked consistently for over 100-years and are very rare to find.

Those aspiring to gain worldwide influence must constantly monitor global conditions to determine any potentials for gain, the strategies promising benefit, and the best times for taking action. Patience must be matched with the ability to seize opportunities and strike like lightning.

The mark of a supreme leader is someone who is always scanning for opportunities. Because opportunities missed are gone forever the greatest leaders are alert to seize whatever chances may appear. For instance, they commonly take action against opponents who show signs of strain or weakness because this type of opportunity is prime for gain.

The saying, "Things must rot before worms may come" aptly

forewarns that many a country (or organization) will first show signs of internal weakness before it enters economic decline. For instance, America's strange emphasis on issues such as white supremacy, destroying its domestic oil industry, uncontrolled immigration, the denial of blatant election fraud and the politicization of the FBI and Department of Justice etcetera all reveal an internal rot and cancer in the national fabric, which usually happens at the end of a supremacy cycle. Other countries will take advantages of these deviances. If a professional sports team loses several key players due to sickness then an opposing team will use this to their advantage, and competing nations will make competitive moves as well using someone's weakness as a time to attack.

If during difficult times you can take over (or simply hurt) an opponent whom everyone else also despises then even your competitors will applaud your actions because they also deplore its existence. Chaos and upheaval are therefore a time of potential gain for the wisest individuals— whether they are statesmen, military officers, business executives, or ordinary men—and this advice to make use of difficulties accordingly applies to every area of life. Some countries will even secret foment internal unrest in others to weaken them. For instance, Arnold Schwarzenegger said that he would psychologically unbalance other bodybuilders on purpose during competitions and then try to capitalize on their unease to help win.

Do you not think that other countries are trying to unbalance others as well such as by giving them wrong economic or financial advice, sending drugs across their borders to weaken them or immigrants to drain their resources in social welfare spending, financing non-profits that will erode their national fabric or cause civil unrest, or emptying their own jails of criminals and sending them into competitors so that they become a problem?

Rather than face chaos and change with fear a political leader should calmly and clearly evaluate whether any dangerous situations he experiences offer any chances for gain. Be careful though because fire spreads from house to house and sickness is transmitted from community to community so another's problems can also become your own if you are not careful enough to protect yourself. In any situation the wisest individuals evaluate the risks they face and then protect themselves by taking appropriate precautions.

An oak tree comes from a tiny acorn. Whatever is big in time always starts from humble origins. You should therefore strive to prevent, or be quick to fix any small troubles that appear in your realm of activity before they can develop into even larger difficulties. If you don't extinguish the smallest flames, such as a banking or debt crisis, they will eventually grow into a large conflagration later on. Martin Armstrong, from his many decades of experience working with central bankers, always says that the problem with most politicians is that they never act to prevent crises because they would rather one occurs so that they can say they will get the guy who caused it and thereby win re-election. You cannot be this way if you want your country to experience prosperity, which is the target rather than your own political power. The supreme political leader of the world does not have a weak country.

The person who excels at eliminating the misfortunes of the people is essentially managing them before they ever appear. "Do the great while it's still small and the most difficult thing while it's still easy." This will make you the most successful of men. The sooner you take care of something the less energy you will need to eliminate any consequential complications.

Master Guan teaches that "Fortune is hidden in misfortune, and misfortune is hidden in fortune" so a wise leader should take advantage of the difficulties befalling other camps to achieve any of his grand objectives. Although he may feel safe during periods of good fortune he should always be wary of potential problems that have not yet become manifest. The wisest of the national saviors are those who remain keenly alert to the factors that may cause their nation's downfall. They fend them off before they ever arise or cut them off in their initial stages so that they cannot grow strong roots but are destroyed. They observe their neighbors' conditions and policies, reflect upon where they will lead, and then abandon their own affairs that might also cause harm.

They also adopt, promote and bolster whatever will lead to prosperity. They make these decisions based on history, their knowledge of human behavior, and by viewing the current state of the world and the actions of its players in their jockeying for power and predominance.

LESSON 6: Leaders Must Actively Demonstrate Their Virtues

> Those who wish to become the ruler of a realm must first demonstrate their virtue to the other feudal lords. Therefore the former kings knew when to take and when to give, when to yield and when to be uncompromising as the occasion might call for. As a result, they were able to become rulers over the entire world.

You must first openly exhibit fairness, generosity and ability if you wish to become the leader of a group. Everyone should know your character and your potential for effectiveness. As to fairness, even criminal organizations establish codes of conduct for their members. Strangely enough, it is because the heads of criminal organizations adhere to such codes that they can maintain control over their (unethical) members.

To become a leader of an organization you will gain support by proving you have adequate qualifications, which requires that you show ability and character. Those who gain leadership positions by force, which simply demonstrates power, differ markedly from those who rise through "virtue" and ability.

Men and women who must submit to others because of their greater force never truly pledge their hearts to their superiors. The only way to win them over is through benevolence. As a result, your abilities and personality (character) must become known and approved by the people you wish to lead; you must become an individual dependably known for ability and goodness in order to elicit a following of loyalty and commitment.

This is important when you lead an organization or state, but *essential* if you want to attain the position of leading the world. You must become a character in everyone's mind that they all recognize for virtuous traits in order to win their acceptance. For instance, the *Army Field Manual* describes the core character strengths of a leader as loyalty, duty, respect, selfless service, honor, integrity and personal courage. If you lack such virtues then it will be impossible to gain the helm. Historian Rufus Fears often stated you must also develop a bedrock of principles that you live by and a strong moral compass to become a political leader, and people must come to know this about you.

People respect all sorts of virtues such as honesty, reliability, wisdom and confidence but to lead a group of people it is especially important to be magnanimous in addition to fairness. Generosity has always been one of the top ways to extend one's influence because your largesse affects people directly and they then feel beholden to you. No one should under-estimate the power of generosity and its ability to win loyalty. A story from China's Sui Dynasty aptly illustrates this point.

When China's short-lived Sui dynasty was overthrown a civil war broke out as several lords competed for power. After a particularly brutal struggle Lord Li Yuan won the fighting and became the new Emperor Gaozu founding the Tang dynasty after which he collected all the booty of the struggle and confiscated all of his opponents' valuables. When his senior generals found out that he was intending to distribute this immense booty to the soldiers and the people they suggested that he treat the ordinary citizens and slaves differently.

As the story goes, the new emperor responded to his advisors: "Why [should I be generous in giving for all groups]? The enemy's arrow and swords had no eyes and didn't favor either of the groups. Taking the same chances, the slaves also risked their necks for the victory and the reward. It's only fair that I treat them equally." He then gave out petty positions with salaries that would support the recipients for the rest of their natural lives to those soldiers who wanted to resign and return to their native villages and become farmers again.

This generous behavior was unheard of. All the advisors, treasurers and accountants were astonished. They vigorously protested to him about this policy and its potentially huge costs for the national budget in the future.

The lord replied, "Don't be ridiculous! I want to be a emperor. How can I afford to be stingy? A penny-wise person will never build an empire. One of the former dynasty's biggest mistakes was its stinginess. The rich had accumulated tremendous wealth and didn't spare anything for the penniless, who rose up and defeated them. Be farsighted and look at the whole picture. I am only one of a handful of potential rulers. To secure my position I must win over the people, not with force and battles, but with money and positions. Politics is an art, not a science. It has nothing to do with accounting and balance sheets. Why should I not be generous?"[11]

Cyrus the Great also won his empire because he was extraordinarily generous to the lords who fought for him. He also told his close advisors that it did not matter that he gave them extraordinary gifts and riches because he was trying to win an empire and was storing his wealth in their allegiance to him. He won their loyalty through his generosity. A pertinent story from Xenophan's *Cyrus the Great* is as follows:

> One day at a dinner in Babylon, Croesus criticized me for being so openhanded (in generosity to others), claiming that I would soon reduce myself to poverty.
>
> I replied, "How much wealth do you think I could have amassed already, if I hadn't shared the spoils of my empire with my friends?"
>
> Croesus named an enormous sum.
>
> I said, "Croesus, I'm going to use Hestifer here to put your theory to the test." Turning to Hestifer, I said, "Go around to my richest friends and tell them I need money for a great enterprise. Have each of them write down the amount he can give me. Then seal their letters and, when you get back, come to me with Croesus and, in my presence, hand the letters over to him."
>
> On the next day, Croesus lost all the color in his face when he totaled up the amount that my friends had pledged. Gasping, he exclaimed, "This sum is so much larger than anyone could ever keep for himself!"
>
> "You see, Croesus," I pointed out, "I do possess common

---

[11] *Wisdoms Way: 101 Tales of Chinese Wit*, trans. by Walton Lee, (YMAA Publication Center, Jamaica Plain, Massachusetts, 1997), p. 11.

sense, you've advised me to harvest and hide it – and be envied and hated because of it, and hire mercenaries to keep an everlasting watch over it. But I've decided to make my friends rich, and they've become living treasuries for me, and they're better at guarding their gold than any watchman could ever be." "And that is exactly why," the downcast Croesus said, "you've won half the world and I've won no more than permission to sit at your table." He shook his head.[12]

Despite the importance of generosity we must recognize that being generous or gracious does not mean being soft. It also does not mean you should abandon sound and just principles in order to climb the ranks of power because a true leader stands firm in his convictions and, as Rufus Fears stated, never deviates from the principles of righteousness. This is why Lord Li Yuan treated slave and commoner alike. Ultimately, your character and deeds are what will win you the respect of your peers, who will or will not accordingly take you as their leader.

Becoming a leader therefore requires high ethical standards that adhere to an admirable code of conduct. It requires the ability to know when to be strict and when to be lenient, when to give and when to take, when to be generous or not, and when and what to say as well as when and what not to say. The ideal mentality represents a balance of firmness and flexibility wherever appropriate.

Actions that are principled, fair, just and balanced are respected by all people and can be continued almost indefinitely because they will elicit approval and even grudging respect from those who may dislike you. Because of these factors, you must never succumb to the notion that power is the only requirement for becoming a supreme leader. Your character and behavior are important in order to hold onto your position.

There are two excellent stories, which involve the outstanding Chinese military and political strategist Zhuge Liang, that illustrate these points. When Liu Bei of the Chinese state of Shu was trying to become the emperor of China the strategist Zhuge Liang served as both his prime minister and as a powerful general in his army. At one point in time Zhuge Liang was guarding some mountain passes with a small contingent of troops who were due to be replaced regularly. Suddenly an enemy state decided to attack Zhuge Liang's position with forces that outnumbered his own troops by almost three to one, and this overwhelming danger just happened to coincide with a time when Zhuge Liang's troops were due to be replaced via their normal rotation.

---

[12] *Xenophon's Cyrus the Great: The Arts of Leadership and War, Xenophon*, ed. by Larry Hedrick, (Truman Talley Books, New York, 2006), pp. 260-261.

Because of this great danger Zhuge Liang's aides urged him to retain the departing troops another month in order to bolster the strength of his forces but he refused their advice saying, "My military command over our soldiers is based on trust and good faith. To lose people's trust by trying to gain an advantage over them is a mistake made by many. Remember that those soldiers whose service is finished always pack their things quickly and wait for the time when they can finally leave while their wives and children anxiously stand waiting back home counting the days until their return. Although we are presently facing a crisis, we cannot abandon what is right and just." Thus, he refused to cancel the discharges and urged all those whose tour of duty was done to leave for home.

When those soldiers due to leave heard these remarks, they were so grateful for Zhuge Liang's kindness that they agreed to stay for one more battle, saying to one another, "Even if we die it's still not enough to repay Master Zhuge's kindness." Because of such high morale, on the day of battle the outnumbered troops rushed forward and won a great victory. In analyzing this victory we must attribute its success to Zhuge Liang's virtue (principles), as this is what won over the good faith of his men.

At another time Zhuge Liang personally led his troops against some warring southern tribal territories that were ruled by a King Menghuo. Knowing that he could easily conquer the region militarily but that there would be revolt as soon as he pulled out, Zhuge Liang realized that he must win the hearts of the southern people so that there would be no future trouble in the region. Six times he captured King Menghuo and his men and six times he showed them kindness and hospitality before releasing them with the admonition not to fight any more. While the enemy troops sincerely appreciated Zhuge Liang's kindness—including the fact that he had spared their lives—King Menghuo would not yield even though he claimed upon each capture that he would finally accept defeat and acknowledge his fealty if captured once more.

On the seventh battle encounter Zhuge Liang's great humanity showed through most clearly. While surveying the battlefield where he had scored a tremendous victory over the southern tribal forces he lamented the terrible cost of human lives suffered by his enemy! When King Menghuo was brought before him this seventh time and was just about to be released again this arrogant tribal ruler finally broke down in tears crying, "Seven times I've been captured and seven times released. Nothing like this has ever happened before. Although I am of the southern people and do not share your culture it doesn't mean I lack a sense of what is right and proper. How could I be so shameless!"

So saying, King Menghuo and his party fell on their knees, bared their torsos in a sign of penitence, and crawled to Zhuge Liang's tent where they recalled their opponent's past mercy and finally pledged their fealty to his state of Shu. In return, Zhuge Liang once again treated them all with great kindness and, surprisingly, confirmed Menghuo's continuing kingship while restoring the conquered territories to his rule. From that point on, there was finally peace on Shu's southern border. It can be said that the tribes could never be defeated by military force, but were actually conquered and transformed by Zhuge Liang's virtuous behavior.

In Chinese history there were perhaps a dozen or so great men like Zhuge Liang who did not exhibit selfishness while occupying great positions of political or military power. Rather, they used their positions for the benefit of the people. The reason Zhuge Liang's stories aptly illustrate Master Guan's lessons is because he demonstrated the inherent principle of virtuous leadership, and many of his successes are often attributed to his morality and high ethical standards.

Because of his virtuous character, Zhuge Liang was not only able to score victory in battles but also win the hearts of the southern tribes. As testimony to his greatness, King Menghuo's people not only recognized the leadership of Shu willingly but they even built a shrine in Zhuge Liang's honor!

As seen in this last example, possessing superior military capabilities is not always sufficient for becoming a supreme leader. Power is not the only requirement for success. You only attain real power and influence when the people under you submit with their hearts and willingly provide their support and allegiance because of the respect they hold for you as an individual. Men will only respect you because of your character or virtues.

Cyrus the Great, for instance, was the founder of the first Persian (Achaemenid) empire but is recognized for his achievements not just in military strategy and politics but *human rights*. Like Zhuge Liang he was distinguished equally as a statesman and soldier but recognized for something beyond. His achievements far outshine those of Alexander the Great because the stable administrative infrastructure he developed for the Achaemenid Empire lasted well past his death. The Babylonians regarded him as "The Liberator" while the Iranians regarded him as "The Father." On this the historian Xenophan commented,

> And those who were subject to him, he treated with esteem and regard, as if they were his own children, while his subjects themselves respected Cyrus as their 'Father' ... What other man but 'Cyrus,' after having overturned an empire, ever died with the title of "The Father" from the people whom he had brought under his power? For it is plain fact that this is a name for one that bestows, rather than for one that takes away!

The task of achieving ultimate world leadership (whose non-political equivalent is the leadership helm of a corporation or other organization) therefore necessitates winning the respect of all the world's citizens because of virtuous behavior, and a position of respect cannot be obtained simply because you possess military, economic or political might. When others recognize that you are honest, fair and balanced as well as concerned about the welfare of everyone then the opposition they would normally hold will often fall by the wayside.

LESSON 7: Territorial Strength

> Military strength flourishes on the basis of power and power is contingent upon territory. Therefore, when a feudal lord enjoys the advantage of territory then power will follow. If he loses the advantage of territory, however, his power will diminish.

For a country to become a superpower it requires wealth and great military strength. You need sufficient military means to take offensive actions and defensive actions to protect yourself when necessary. Sima Yi's five principles of warfare are to fight if you can fight, defend if you cannot fight, flee if you cannot defend, and the remaining two options are either to surrender or die.

In earlier times, the strength of a nation was clearly related to its geographical size, terrain and population. The size of the population equated with potential soldiers and farmers, and the number of farmers equated with agricultural production and national wealth. Some countries were simply gifted with good or bad geographical circumstances and would invade others to enlarge their realms. The decisions a country makes, such as this, are therefore what determine its success over the long term.

A large territory represented a great amount of wealth but only if the geography was fit for agricultural production. Mongolia, for instance, has a large amount of land but it is only suitable for herding sheep or goats rather than farming so it cannot support a large number of people. A large extent of arable land, on the other hand, can produce lots of food and support more soldiers.

The problem with today's fertile land is gradual soil degradation (see David Montgomery's *Dirt*), water availability, clean seeds and rainfall issues. Permaculture can help address many of the water table and regenerative agriculture issues, and the artificial rain-making methods of Trevor Constable (see *Loom of the Future: The Weather Engineering Work of Trevor James Constable*) can help create rain under the right conditions even when nations are using weather warfare against you to produce drought.

I've actually used his (reflective coated ABS plastic tubes with a double-sided mirror on one end) weather tubes that he taught me how to make to produce rain in the Mongolian desert and have many real life stories about their veracity. I've seen their effectiveness with my own eyes where the double-sided mirrors were pointed at the direction of atmospheric energies to cause them to compress together and back up like a traffic jam, form clouds and then discharge rain. You can even connect a car battery to old television antennas in ways that create standing waves that prevent rainfall too, or use Constable's weather tubes to knock out one of the legs of a hurricane or typhoon. National leaders need to know about such methods which can protect their country's agriculture so they can produce rainfall during natural or artificial droughts and not be blackmailed by other nations who can control the weather (or earthquakes) with energy technologies such as HAARP. These methods are even employed after successful invasions to further cripple the economies of nations. The threat of using directed energy weapons to destroy a national economy has been used to blackmail many nations hence such responses should be known.

Economies today are much more developed and diversified than in earlier times. International trade is so efficient that even smaller nations, such as a Japan, can become superpowers in spite of their size and lack of resources if they position themselves properly in the economic sphere. The modern correspondence to "territory" is therefore best equated with the national collection of "productive assets," which includes a nation's people along with its agricultural, industrial, financial and commercial base.

In Master Guan's view, and taking all other things as being equal, nations with the best workers and possessing the strongest agricultural, commercial, industrial and capital bases as well as trade flows can become the most powerful in the world. However, while the power of a nation will be proportional to the quality of its people and its productive capabilities it is simply not enough for a nation to have productive assets. The assets of a state must be put to use in order that it can gain economic and then political power.

When the economy of a nation is experiencing a recession, depression or some other form of economic crisis then its territorial strength will naturally diminish. The economic problem that most often afflicts nations involves manipulations of the money supply leading to excessive debts and then subsequent boom and bust cycles. Carmen Reinhart and Kenneth Rogoff wrote a policy guide for political leaders in *This Time It's Different* which addresses internal defaults, external defaults, banking crises, exchange rate crises and inflation crises.

A strong industrial and commercial base will support wide broadscale employment whereas if this base is lacking the potential for economic prosperity is meaningless. Cultivating a strong commercial foundation is the how you insure a variety of economic goods and services that will increase national revenue and wealth. In an individual firm, commercial success and prosperity depend upon offering a desirable set of products and services for sale to others and when you scale this up across cities and then the nation you arrive at national prosperity or not.

The key to a nation's strength, power and standing is to possess revenue producing industries and assets. If those assets are destroyed or left idle then a nation will lose the means to feed and defend itself. This is why it is important for countries to create entrepreneurial environments which encourage agriculture, business, commerce and investment.[13]

When a company starts losing product sales and market share in the commercial field this is the same thing as having its life force drain away. When a nation's sources of generating wealth similarly decline and the country loses its productive assets then that nation will also experience decline. In the field of geopolitics it is easy to understand how countries whose economies hinge upon the declining reserves of a small set of products[14] will face poverty unless they take steps to diversify their economies.

History teaches that no set of conditions remains stable forever. A nation can lose its wealth and the means to defend itself and then face ruin because no nation has been granted the boon of perpetuity. Nations must strive to build and maintain strong economies because a flourishing economy is the only way for a nation to maintain both its political influence and military might. Since the countries of the world are often treated by each other like customers in a gambling casino where the wealthy are pampered and the poor are ignored it is crucial to become an economically prosperous nation. Nations should therefore become devoted to strengthening their economies because the wealth boost will help them increase and project their worldly standing and influence. This is a fundamental principle to geopolitics.

Furthermore, it is hard to promote social stability and ethics in societies where the people are poor just as it is hard to teach virtue when the people are hungry.[15] When the people are prosperous, however, a

---

[13] This is why countries should pay special attention to their tax codes, which are a main factor behind investment incentives. The tax policy and regulatory code for a country can be a major source of investment deterrents or incentives.

[14] Saudi Arabia (a supplier of oil) or the island of Nauru (a supplier of phosphate) are examples that come to mind. Saudi Arabia has designs to switch from being a petroleum producer to mineral producer because its oil reserves are running out. In many places the U.S. is rapidly depleting, through destructive use, its water and soil resources, which bodes ill for its long-term ability to feed itself.

[15] As an adjunct observation, Walter Bagehot (the early editor of *The Economist*

domain can be made orderly whereas an impoverished people become desirous and disaffected and easily seduced by profit for survival's sake. People will always risk breaking the law and suffering the consequential punishments if they are hungry and cold. Afterwards you can talk to them about ethics. The means to reduce the crime rate is to increase a nation's prosperity and employment.

LESSON 8: Leaders Always Need Public Support

Those who would contend for control of the realm must first win the support of the people. Those who understand the grand scheme of things will obtain the people's support while those who are short-sighted will (act in ways that) lose their support. Those who succeed in winning over the masses of the realm become kings while those who win only half of the support become mere lord protectors.

While a country needs economic and military might to become a superpower an individual needs to win men's allegiance in order to become their leader. Individuals who really understand the art of power recognize willing support (a loyal following) as the crucial requirement for leadership. If you have no followers then you certainly aren't anyone's leader.

Since rulership is built upon the foundation of the people's will no true political leader can long exist without the public's support. Any individual aspiring to become a supreme leader must therefore strive to capture the majority of human hearts just as politicians attempt when they go out on the campaign trail.

Confucius, like Master Guan, spoke about this great need for winning over the people. Confucius said:

He who gains the hearts of the people will gain the kingdom, while he who loses the hearts of the people will lose the kingdom. Hence a ruler should first take pains about his own virtue. Possessing moral virtue, he will gain the people; Gaining the people, he will obtain territory; Possessing territory, he will obtain revenue and wealth; Having wealth, he will possess the resources for expenditure and have the means to act. Therefore,

---

magazine) once noted, "All people are most credulous when they are most happy." Hence it is easier to retain the leadership helm when you successfully bring about a condition of prosperity. Officials therefore try to create an economic boom just prior to an upcoming election.

virtue is the root and wealth is the consequence. If a ruler disregards the root and esteems the consequences he will only wrangle with his people and teach them to be rapacious. Thus, if a ruler thinks only of amassing wealth the people will disperse from him, but if he disperses wealth among the people they will mass round him. And hence, a ruler's words and decrees that are contrary and unjust will be met by words that are rebellious in the same manner; and wealth that is ill-gotten or amassed by improper ways will be lost in the same manner.

This confirms many of Master Guan's teachings, especially the principle that gaining the people is accomplished by cultivating your character. An example is President Dwight Eisenhower who cultivated a sunny optimism (he was nicknamed "Sunny Jim" at West Point), the knack of saying the right thing to gain the cooperation of other people, and whose good nature inspired the trust of others. He actually set out to develop these traits in himself.

Confucius and Master Guan both agree that self-cultivation like this is the fundamental root source behind the ability to attain power and influence. Self-cultivation results in a more likeable personality so that you naturally gain readier acceptance by others.

Today we call self-cultivation the principle of "character building" and "knowing your own mind." It means working on decreasing your rough edges and personality flaws, and always reflecting upon yourself and your actions so that you are continually in a mode of self-correction and self-perfection. This is how people stay centered in maintaining their sincerity of purpose.

This type of alert awareness prevents you from becoming caught up in a situation where you lose the proper perspective, and it prevents you from going too far with your actions (hubris) or becoming the unwitting tool of forces beyond your control. It prevents your emotions from controlling you and causing you to make mistakes. Introspection is the basis of personal cultivation and success in your personal self-cultivation leads to the ability to smoothly run a family, or business or country. Confucius said that you must first cultivate yourself and this would enable you to rectify your family, and from families being rectified a state would then become rightly governed.

There is an inherent commonality behind Master Guan's idea of the Lord Protector, the Bodhisattva ideal in Buddhism, Confucius' idea of the perfect ruler, and the undefeatable Chakravartin (a "Wheel-Turning King" who conquers everywhere) described in Buddhism and Hinduism. The goal of Confucianism is to cultivate yourself (your thinking, personality and behavior) so that you can properly govern your family, community and country where the ultimate purpose is to bring peace and harmony to the world. A Chakravartin brings peace and prosperity everywhere as well. This is also the objective of a supreme leader and the goal of the Confucian sage king.

Master Guan, Confucius and Buddha all say that a great leader is one who loves humanity, is concerned with its well-being, and cultivates the desire to nourish all the people of the earth. This means universal concern for the welfare and protection of all people.

"Character" (personal virtue) is an important leadership because your conduct and actions are essentially your character expressed. In fact, the crucial difference between someone who becomes the leader of a country and one who remains a minor politician is the scope to which their character extends its caring nature or compassion. Everything is done for the benefit of the people rather than for personal selfishness.

While a great leader sacrifices himself for the whole group a minor leader establishes limits to his benevolence. The great leader is blind to prejudice and class distinctions while the minor leader emphasizes the differences between people. The great leader practices universal acceptance and forbearance. He wins the public's trust through the excellence of his character and ability rather than by simply making promises to help people. However, to become a supreme leader he should promote a grand vision in tune with the nation's core values or DNA to obtain adherents who also want the country to develop in those directions.

Public support leads to power and from power comes the ability to marshal wealth and to carry out one's plans so the fundamental method for becoming a supreme leader is not by engaging in political tricks, saying what the polls show is popular, and relying on false public relations campaigns or other manipulations. The method rests squarely on genuinely cultivating one's character, actions and behavior. You can read more about this in *Color Me Confucius* and *Correcting Zen*.

The path of everyone's life can be traced back to their character because character is destiny. No matter what you promise in order to elicit a following you won't inspire the trust of others if you are seen as dishonest, undependable and lacking in character.

Turning to the commercial arena, what distinguishes dominant companies from the "me too" and "also-ran" is market share, which is the business equivalent to public opinion. A company is similar to a politician in the way it gains market share because it must also publicize its virtues—namely, its reputation for providing quality goods and services—to win public support (product sales). While an individual who gains the hearts of a community can become a political leader it is the company that gains the most market share that naturally becomes the leader of its field.

Political leaders who try to capture the heart of their nation ascend to power when their popular support reaches to all corners of the country, and then they gain real political power. Those who fail in this mission are those who address only parochial concerns, and it is because they limit the extent of their benevolence that they never advance to first rank importance. You must address both the public and the elites who control most matters behind the scenes.

Companies that concentrate on filling tiny market niches, rather than satisfying broad market needs, must also be considered in this light. While they can be quite successful and establish a secure livelihood by cultivating a niche, a niche strategy (by its very definition) will never win the majority of a market. However, a niche offers a quite defensible foothold for entering a market and from which ultimate dominance can gradually grow. From a first step entry point can come an advancement in sales, and then more and more. As a Taoist would say, from one comes two, from two comes three and from three comes ten thousand other things.

As to the various means available for politically winning over a populace there are several methods which have been ever popular since ancient times. For instance, in the days of kings and empires the contender for a throne might declare his legitimate right to rule based on the fact he was the descendent of some well-known predecessor, or a distant relative of the imperial house,[16] and people continue to use this same strategy today. They may stress, for instance, the legacy of a family name (George Bush Jr.) which often assumes greater importance in elections than a person's actual governing capabilities. As government elections over the world have shown—whether in the East or West—the public will often entrust power to a famous namesake even if the individual lacks the basic qualifications for governing.

There is no doubt that a famous name often compensates for a lack in leadership experience. Whether we are talking about the managerial succession in today's public corporations or the election of heads of state, people still correlate legitimacy of office with bloodlines and family ties. There are "royal lineages" within every country, even democracies such as the United States.[17] Those lineages help their bloodlines get into power with examples being President Carter (son of a Kennedy) and President Clinton (son of a Rockefeller). Bearing one of these famous family names

[16] This is the traditional path of leadership inheritance by lineage whereby the oldest male offspring (or similar) of the current leader becomes the new head of the organization.

[17] Examples might include the Kennedys or Rockefellers, which are known as the aristocracies of America. The Bushes, Clintons, Romneys, Pelosis and several other political families are even now trying to establish themselves in positions of continual dominance.

will often open doors which might be closed in other circumstances.

In the past, whenever a contender for a kingdom was unable to claim a royal relationship himself he might falsely set up as leader some powerless individual who possessed that status. After he succeeded in establishing a false ruler he could manipulate the puppet's actions while building up his own support by appointing or befriending established ambitious ministers. This strategy involved controlling affairs quietly until the usurper was strong enough to come out into the open.

The Chinese general and chancellor Cao Cao of the Eastern Han dynasty (during the Three Kingdoms period) used this strategy by controlling the last Han Emperor Xian. The Japanese also used this strategy by establishing Pu Yi (the "last emperor" of China) as the puppet ruler of Manchuria while they maintained behind-the-scenes control of the region. This is a disguised means of coming into power through brute force, but is often accepted by a local population.

While the exceptional leader uses legitimate means to attain power he is never blind to the scheming machinations others use to secure a public following. You must remember that Master Guan could never have lasted over forty years in politics if he was ignorant of the wide variety of stratagems used by political opponents who can be cruel, dishonest and treat others with callousness. Some people by nature have manipulative orientations where they only see people as objects to be exploited or rivals to be defeated; many sociopaths and psychopaths seek positions of power. Master Guan always stressed that a ruler should keep his eyes open and know things by his wisdom, including the evil in men's hearts, their secret designs and methods by which others might manipulate the public to achieve their support.

To be able to remedy malignant conditions you must understand how those conditions operate. For instance, to stop thievery and cheating within casinos the security office needs to know how the various cons operate. Hence, expounding upon this shadowy side of worldly affairs must not be taken as an endorsement for emulating such methods. Rather, alerting you about these methods serves the purpose of revealing hidden workings so that you can be aware of the devilry around you and know how to protect yourself. When you awaken to this sort of revelation you can then start to formulate appropriate defenses to such machinations. For instance, when you see a leader making unrealistic but rosy promises directed to the poor and homeless, the young and unemployed, the discontented and destitute—those who hold no fear of punishment since they have nothing to lose—then you may be witnessing a form of manipulation which is preamble to a revolution.

In any case, Master Guan's lesson is that the individual who wishes to become leader of a realm must put a lot of effort into winning over the masses. He does so by displaying his personal virtues and governing by benevolent means. If the masses are experiencing troubles and difficulties he will win their support by voicing their concerns and promising to address their needs. Words and actions that vibrate in harmony with the people's will are those which will gain their support. You are most likely to win the backing of a nation when you authentically address the needs of the times.

Aspiring leaders must therefore promulgate a grand vision that all can understand. The vision should advertise the values and the principles for which they stand that addresses the needs of the time, as just stated. A leader must constantly communicate this vision to the public through the media and symbolically embody that direction in himself and his life if he can. He should become an idol of beneficence and benevolence.

A leadership position is only as strong as one's support so when an individual's vision touches a majority of hearts he will be in a position to move a nation. However, those concerned with just small issues will only be able to attain a secondary role in the leadership hierarchy.

To become a world leader a nation too must be concerned with others rather than just itself. It needs to pursue and promote solutions to issues that help everyone instead of itself alone. In the international scene, when a nation wishing to lead others speaks of everyone's common interests then all will listen, but when it only speaks of its own concerns it will usually just alienate its peers.

LESSON 9: Winning Political Support

For this reason the sage kings were humble in order to attract the support of worthy men and employ them in office. They established equitable legislation in order to attract people from all over the realm and distributed their wealth in order to win over the people. Thus they were honored as sons of Heaven and their wealth encompassed the entire realm. When they meted out punishments no one considered them greedy because they were (seen as being) devoted to great plans for the country. The sage kings used the resources of the realm to benefit the people, and used their wisdom and might to bring together the political power of the realm. By acting with benevolence and virtue they won the allegiance of the feudal lords. By punishing the treachery of those who were wicked and vicious they rectified the minds of the people. They made use of military force to stage punitive expeditions all over the country. They attacked rebellious and disorderly states, rewarded the meritorious, honored the worthies and sages, and displayed their own virtues to reassure the common people.

When the former kings were able to gain control of the realm it was because they had mastered such political methods. Of such great virtue were these methods that all things were said to have benefited from them.

Winning the support of the people entails befriending the common man and establishing good relations with the country's top talents. The relationship with the nation's highest tier of talent is important because a good leader must attract the most intelligent and skillful to work for him, who by their very qualifications tend to be extremely selective as to whom they will serve.[18] You must be sensitive to their special needs and interests if you want to obtain these exceptional people. And as Lao Tzu said, "One who excels in employing men acts deferentially to them." He always highly respects them as if they were his teacher.

All nations need a strong civil service just as companies need talented managers. You might be a great manager yourself yet every company also needs skillful employees to put your plans into effect too. Even with wealth in abundance the goals of growing a company or bringing a nation to superpower status cannot materialize unless talented men are put into office. A leader with grand goals therefore needs to attract the best talents available and actively employ them for his benefit.

History shows that the ability to achieve peace and prosperity in a country often rests upon the activities of a few worthy men, so every leader should aspire to put such worthies into office. As Margaret Mead once said, "Never doubt that a small group of thoughtful, committed citizens (such as your advisors) can change the world; indeed, it's the only thing that ever has." Not only will the talented use their skills to help a firm or country grow but their very presence can serve to hold the crafty and corrupt in check. This is why Confucius said, "By raising the straight and setting them over the crooked (the wise leader) can make the crooked straight." Lao Tzu said, "To the good the sage is good, and to the bad he also shows goodness until they too become good." Putting great men into office can help suppress selfishness, callousness and intrigues in lesser men.

The best means for obtaining talent is to make talent come to you,

---

[18] When seeking employment, the exceptionally skillful tend to be attracted to those who accurately recognize their talents because that is the best way to ensure that their special talents will be compensated. When in their presence a leader may think he is the one doing the interviewing, but premier talent may actually be interviewing their prospective "boss."

and one of the best means for doing so is to employ the method of "worshipping the skull." The method of "worshipping the skull" is letting everyone know that you place an extremely high value on talent and broadcasting that message everywhere through stories about how you are willing to greatly compensate talent. Upon hearing this message the talented will automatically come your way without any extra need to seek them out. The best like to work with the best and especially with those who appreciate their efforts, so if you can establish a reputation for honoring the best and being willing to follow their advice then talent will flock your way.

You can think of the phrase "worshipping the skull" as meaning that you treasure a talented individual (the skull), but this strategy is named after the story of a Chinese lord who wished to obtain the best racing horses in China for his stable. Because the country was so large he was at a loss as to how to secure the country's fastest horses unless he could come up with an effective strategy. Finally he thought of a way to do so.

One day he went and purchased the worthless remains of a famous thoroughbred that had died and paid an enormous price for the horse's skull. He then set the skull up in his house's shrine and began to worship it every day. The story of the lord's exorbitant purchase and this outrageous behavior of worshipping the skull spread so quickly that soon individuals from across the land came to offer exceptionally fine horses for sale. They reasoned that if the lord so honored fast horses that he would worship the skeleton of a dead thoroughbred then he would certainly pay top dollar for excellent living specimens.

Confronted with a similar situation, King Zhao of the Chinese state of Yan once despaired that his state was lacking in talented ministers. How could he attract talented ministers so as to build up his state? Noticing his concern, one of his junior officers gave him the following advice: "I have little ability myself, but you can gain the talent you desire by using me as a start. If you treat me as a teacher and build me a fine house, you will quickly establish a reputation for valuing talent that will attract the worthiest people to your employ."

King Zhao quickly adopted this strategy and built his advisor a magnificent mansion and thereafter treated him openly with great respect. In a short time the king's reputation for honoring his advisors spread far and wide, which succeeded in attracting the desired worthies to his state.

When you establish a reputation for honoring talent with large rewards, as many Arab elites have done with their riches, you will easily be able to attract the best people in the world to work for you. It is your reputation for good or bad, for honesty or dishonesty, and for rewarding people or stinginess that will (among other things) attract or repel people. You must motivate your people by recognizing merit and then the talent will vie for positions to serve you.

Businesses that have the reputation of paying the highest salaries attract the interest of the most talented and most skillful because of this same principle. In fact, the premier firms rightly consider high salaries an investment rather than an expense. Premier firms recognize that "when you pay peanuts you only attract monkeys" so they pay large salaries to employ the very best possible.

In order to attract the elite into civil service, Emperor Wu of the Han dynasty set up a recruitment system that allowed officials of all levels to recommend outstanding individuals for apprenticeships or scholarships to serve the government. This provided opportunities for capable people with less privileged status to work in the administrative system even if they had poor backgrounds. In this way he was able to bring many talents into the government.

After a leader attracts the talents he fervently desires and has them in his employ he must behave like Charlemagne and treat them as his teachers, consult them in his affairs, and never fail to show them hospitality and respect.[19] When a leader shows the talented both honor and respect, they will appreciate such kindness by in turn offering their best work.

When you secure the employment of talented individuals you should do as Duke Huan did and promote the worthy with a view to the nation's future regardless of your personal feelings. This is why Duke Huan took Guan Zhong as his prime minister even though Guan Zhong had once tried to kill him.

Whether individuals are Republican or Democrat, Islamic or Jewish, male or female, gay or heterosexual, if they are talented and trustworthy you should use them to help build the state. This is the principle for selecting individuals to build a nation—use the talented! Race, sex, or color are not important because the only consideration is to find and use the highly talented. You should not try to focus on diversity but promote a meritocracy of the talented.

A king should employ his advisors with the goal of making his country strong and prosperous. This goes without saying. When any riches gained are then used in service of the people then no one will complain if the king becomes wealthier himself. However, if an individual's wealth comes from the people but he doesn't return anything to society then criticism will arise from all quarters of the public. Stockholders, for instance, will criticize a CEO like Disney's Mike Eisner when he doesn't

---

[19] There is an old Chinese maxim on rulership we should note: "To become an emperor you should treat your subordinates as teachers. To become a king, you should treat them as friends. To become a lord, you should treat them as guests. To ruin your country, you should treat them as servants or slaves."

produce profits but they won't say a word about any of his mistakes or shortcomings if he everyone becomes richer while he is in office because the stock price starts to rise.

When people think a man is wealthy but no one envies him it is usually because he has used his money to help others, enriching them through philanthropic and charitable causes that everyone knows about. Hence, when a national leader distributes his own wealth to the people and uses the nation's resources in service of some grand scheme for their benefit then no one will find fault. But if he squanders the nation's resources on personal projects, or diverts its wealth into his own coffers or lavishly spends it on his own enjoyments he will often end up being toppled.

A wise ruler is therefore one who humbles himself in order to gain the support of the learned. He divides his administration among the talented in order to obtain the best and win their support. Hiring the talented is one of the best means for silencing or outmaneuvering your detractors. Only when he finally has good men in his employ can he finally set about putting the realm in order including teaching the public to adopt virtuous ways, encouraging them to discard unethical behavior, punishing the wicked whose actions have been rebellious and disorderly, and toppling cruel tyrants who violate the principles of humanity. President Donald Trump, failed in achieving what was possible for his administration because he ignored this advice and chose the wrong advisors who commonly backstabbed him after he put them in positions of power.

Great leaders never refrain from using force if it can bring peace to the land, and so they will punish criminals because this will in turn protect the people. They're like the great Wheel-Turning Kings mentioned in Buddhism who conquer all nations in order to establish peace everywhere.

In both their rewards to the worthy and in their punishments to the wicked, supreme leaders try to set an example for the public. They teach the people to purify their thoughts and remind them to be responsible for their own behavior and its consequences. They try to arouse a national sense of shame for wrongdoings because many people violate the spirit of the law while staying within its boundaries. If people retain a sense of shame then they can still be rehabilitated because they are not beyond hope. If they lack a sense of shame then it is unlikely that they can be corrected at all. Everyone wants to see a sense of righteousness being displayed in their leader.

When the leader of a state acts with moral righteousness then the people will listen to him. When he has moral righteousness on his side then his military forces will fight with high morale, which is a powerful force that can help win battle engagements. Confucius spoke of giving oneself to justice and Mencius spoke of sacrificing for righteousness, and thus a leader displays these qualities in his actions and intentions.

If a leader exhibits his virtues and strives to set an entire realm in order then he can eventually be made a king. Once he becomes a king he can safely maintain his position if he uses the country's wealth for the people's benefit, for establishing law and order, for comforting and taking care of the suffering, for assisting people in danger, and for befriending the leaders of other states for his own nation's benefit.

When a king rules using virtue and justice as his guide then all the people will benefit from his actions. This is what will make his reputation and standing grow over time.

LESSON 10: The Sage Kings Are the Best Models for Leaders

Those who act to make sure that their states never suffer calamities or troubles, but accrue both fortune and repute, are [the highest rank] enlightened sages. Those who preserve their states facing danger and destruction are [the secondary rank] wise sages. Therefore (to accomplish these aims) the former kings considered the enlightened sages as their teachers and revered them as great saints. A great sage is one whose advice, if heeded, can save a state from destruction and whose advice, when ignored, can result in its downfall.

The individual who wants to become a successful leader must study the lessons of history to avoid the mistakes of the past. A leader must understand the tides of yin and yang that rule events and how to manage these transformations for his nation's benefit. Those who don't reference the past are condemned to repeat its mistakes just as those who don't understand the trends of the times and extremes of yin and yang will tend to act improperly. He must also understand how to wield psychology like a weapon in tune with these trends.

To manage a country the principle of eliminating mistakes is often more important than establishing new initiatives. If you administer a state like weeding a garden to get rid of whatever chokes the vegetables then what is left alone will tend to prosper without anymore suppressive entanglements. This is the reason why laissez faire policies have helped many countries prosper while too many regulations have often produced economic stagnation.

In business there is a "zero defect policy" where you just focus on preventing mistakes and then incredible benefits will flow down the line. In the investment field we say, "Cut the losers quickly and profits will take care of themselves." Many fields of human endeavor emphasize the same principle that decreasing errors is often the best way to promote growth rather than putting fertilizer on a situation.

If we look at history carefully it is easy to see that many rulers failed because they undertook extra initiatives when things were good enough. As George Kennan once observed, a man should choose his battles carefully because some problems should be left alone. Due to excessive desires (and sometimes overconfidence), many rulers of the past succumbed to the temptations of "imperial overstretch"[20] only to see themselves topple because of hubris. Like Napoleon's catastrophic advance into Russia, many leaders failed because they tried to extend their state too far. The list of such cases is endless.

Another typical problem is when political leaders tried to change the internals of their state too quickly or too much. Lao Tzu's advice was that governing a country should be like frying a small fish. If you turn it over too many times both a fish and country will break into pieces, so you should interfere with matters as little as possible. Lao Tzu also said,

> When the government is too intrusive, people lose their spirit;
> Act for the people's benefit; trust them. Leave them alone.

In general it is therefore best that governments refrain from intervening too much in the peoples' private affairs. One particular mistake is for nations to become nanny states that try to over-protect citizens, which happens when a government tries to establish too many regulations. People have to make their own mistakes and learn personal responsibility for their own deeds. A nation should encourage its people to examine their own situations, act accordingly with prudence and to take responsibility for their behavior and its consequences. After all, this is what makes people stronger and wiser and strengthens the national character.

It is also better when governments rely on simple programs and initiatives to get results like using sunshine to melt ice rather than by instituting complicated schemes and policies to attain their goals. Good advisors can develop the appropriate policies for accomplishing government aims, and they can also warn about policies that will probably fail. When considering such policies there is no excuse for not taking into account history and the constancy of human behavior.[21] Successful policies

---

[20] A term coined by historian Paul Kennedy, who identified the trend toward excess as an important toppling factor behind many peoples, powers and governments. The idea of overstretch or overreach is essentially the principle of a yang extreme leading to the birth of yin (decline). Just as you can overeat or indulge in sexual excess to your detriment there is also the possibility of the "impulse to conquer" leading to a detrimental extreme.

are always based on simple principles as to how people usually act.

History and psychology are not the only sources you can reference when you're a leader trying to create good policies because the sages of the past have also bequeathed to us their wisdom. Of particular note are the *Guanzi*, *The Three Strategies of Huang Shigong*, the *T'ai Kung's Six Secret Teachings* (written by the Chinese sage hermit Jiang Ziya)[22] as well as the works of Confucius, Mencius and Laozi, all of which are full of great wisdom for the aspiring executive, administrator and statesman.[23]

Countless are the men who have won kingdoms or become great rulers by following the advice of sages such as these. A lesson we'll see in the *Guanzi* again and again is that a great leader must learn how to judge correct timing. Quite often timing is more important than strategy.

An enlightened sage can readily discern the effect of policies and decrees and has insight into proper timing because he has the ability to know the future. Because of this ability we say that true sages can destroy a nation with but a few thoughts, or break an army through their deliberations. The sages are the foremost individuals to be heeded. In fact, the strategies of the sages are powerful enough to move nations where mighty armies would prove ineffective. They can conceive of brilliant measures whose achievements cannot be blocked.

---

[21] If you subsidize certain behavior you will get more of it. If you punish certain behavior then you will get less of it. All policies should therefore include an aspect of encouragement and discouragement. For instance, if welfare policies are based solely on the policy of giving, giving, and more giving then they are sure to lead to abusive extremes. All policies should be so artfully derived as to take human behavior, sociology, psychology and history into account, and thereby more likely to successfully achieve their objectives.

[22] An excellent single source for several of these works is *The Seven Military Classics of Ancient China*, by Ralph D. Sawyer (San Francisco: Westview Press, 1993). Another translation is *The Six Secret Teachings on the Way of Strategy*, by Ralph Sawyer (Boston: Shambhala, 1996).

[23] Just as a good general should be intimately familiar with the military works of different nations, a good statesman or diplomat should be familiar with the political works of other countries in order to understand the cultural roots behind the variety of rulership concepts in play. Such works would include Kautilya's *Arthasastra*, Machiavelli's *The Prince*, *The Book of Counsel for Viziers and Governors* of Sari Mehmed Pasha the Defterdar, *The Republic*, and so on. A wise statesman should also study the history of rulers who failed and those who succeeded during trying times. Chinese history offers an unbroken stream of such historical records, and there are also the histories of the Roman Empire. Aspiring statesmen might study the policies of China's Tang dynasty, or the period of the "Five Good Emperors" of Roman history [A.D. 96 to 180] to see what lessons they can glean.

Lao Tzu said, "Only sages are effectively able to know strategy, so their words prove truthful and their expectations prove accurate." That's why the wisdom and advice of the sages are the necessary consumption of responsible leaders.

Words express one's inherent wisdom so when an enlightened sage issues a warning or offers you other timely advice it is best to take note. His advice can help save countries from disasters or a nation can experience a downfall when the advice is ignored. Confucius is reported to have said, "A gentleman stands in awe of three things: the will of Heaven, great men, and the words of the sages."

Chinese King Wen and his son King Wu, who are reputed to be the authors of the *I Ching*, owed much of their administrative and military success to the advice of the sage Jiang Shang just as Duke Huan succeeded primarily due to Guan Zhong's guidance.

The ancient sages who provided such advice were not idle philosophers or untested theoreticians but were extremely accomplished in practical affairs. Their exceptional insight and talent owed their origins to self-cultivation, and they became known as "sages" because they used the fruits of their attainments to help assist and guide others. By no means were they men lacking in practicality as we often find with the liberal columnists, consultants and so-called "experts" surrounding our capitals today. They were men of wisdom, insight, experience and ability—men who understood practical affairs and human behavior. The teachings of the sages carries the weight it does not only because of their transcendental wisdom but because they were practically skilled in worldly affairs.

Confucius, for instance, first worked as a humble clerk in a granary, and later rose to become an overseer of flocks and grazing grounds. He next worked as a foreman of building construction, a steward, and finally as a governor of a district in the Chinese state of Lu. His later experience was even broader than that of Theodore Roosevelt because it ranged through a variety of offices such as Minister of Public Works, Police Commissioner, and Minister of Justice. So able was his administration that China's other territories feared that Lu would become China's most powerful state because it employed Confucius.

All leaders should study the writings of the enlightened sages and sage kings who managed their countries to avoid any major difficulties. Those who can prevent their countries from experiencing major troubles should be considered enlightened rulers. Those who merely save their countries from the brink of danger deserve secondary status since the enlightened sages make sure no trouble arises at all. This is a capability the Japanese seek when selecting a corporate chairman. The Japanese say it should seem as if he is doing nothing at all as his company prospers and it by this means that you can know that he is doing a great job.

Whether we are talking about nations or commercial enterprises, those individuals who wish to become powerful leaders should learn the wisdom of the sages. The sage whose writings are most relevant to us today is Guan Tzu, or Master Guan.

LESSON 11: Leaders Must Establish Priorities

> The wise king attaches little importance to horses and precious stones [treasures]. Rather, he values good government administration and a strong army. The incompetent ruler worries about just the opposite. He is casual in conferring government posts, but is scrupulous when it comes to assigning horses. He is lax when it comes to staffing his army but diligent in bestowing gifts. He pays attention to the guardhouses of his palace gates, but neglects the defense of his country's borders. For these reasons his territory will be pared (whittled) away.

This passage can be translated in many different ways but in each way there are valuable lessons to be learned.

First of all, it's important that political leaders who wish to govern well should economize their budgets, watch their expenditures and avoid as much debt as possible (where one of the biggest debt drivers is war). This will conserve state funds, lessen the burden on the public and reduce the nation's risk of ruin because countless countries in the past have collapsed due to their debt burdens. It is common for social welfare spending in a nation to increase over time which then requires more debt and taxes that are extracted from productive private industry and its commercial development effort. This takes away the capital necessary for new innovations, impedes a nation's industrial growth and leads to economic decline.

Political leaders should also be frugal in their own personal affairs by avoiding extravagance in order to set a positive example for the public. The leaders of the past who squandered state funds on luxuries and palaces rather than on strengthening their governments often bankrupted the state due to overspending.

It is essential that the resources of the realm be used for basic concerns—such as military defense, economic development and social welfare—rather than diverted toward the superficial interests of the ruler. The best investment is in the people's welfare. Furthermore, the *I Ching* advises, "Those above secure their homes by kindness to those below."

Modern times are in many ways the same as the old for today our government officials don't directly dip into the government till for themselves but use the public funds to sponsor special interest projects that profit their friends who then reward them. This is no different than amassing personal wealth by stealing from the public. As more and more officials adorn themselves through this policy it eventually overloads a nation with corruption.

It is impossible to stave off ruin when a nation's leadership increasingly enriches itself through bribery or by siphoning off funds from the public purse for useless projects that provide kickbacks. The very art of governing means bringing the people out of poverty instead of increasing their hardships and burdens through more laws and taxes to make up for mismanagement. If a government feels that taxation to fill deficits is the way to bring people out of poverty then it is heading down the wrong trail.

In economic affairs the best banking policy has always been to store the nation's wealth among the people rather than in the government. This is accomplished through wise policies that restrain government spending, but only wise and prudent officials could ever implement such policies because most would rather see what they could get for themselves.

Good rulers will therefore stress good government and military policies while mediocre rulers will stress their personal desires. Master Guan states that those countries whose leadership helm worries more about its personal affairs than matters of state will see decline over time.

LESSON 12: Leaders Must Cultivate Power, Wisdom, and Their Own Independent Thinking

Political power is a major asset to an enlightened sage. His unique wisdom is the sharpest weapon within his country. His independent decision-making is the same as his nation possessing secret defenses. These three are what the sage relies upon. The sage senses danger when it is still incipient, while the fool becomes alarmed only when it is obvious to all. The sage hates what is ugly in men's hearts while the fool only hates outer appearances. The sage will discern if it is dangerous before he makes any move while the stupid [unprepared] person, upon encountering danger, wavers and gives way. The sage therefore seizes his chances by taking advantage of the times but he never acts against the times. Those who are merely clever may excel at strategizing but they are inferior to those who make skillful use of timing. Those who are skilled in the art of timing can accomplish much more with less time and trouble.

Political leaders need power, wisdom and the skill of independent decision-making to successfully accomplish their endeavors. These three qualities are the maxims for a king.

Without power—which marshals both men and wealth—there will be no resources enabling you to act.

Without wisdom your plans will fail to accomplish your aims. Wisdom distinguishes leaders from the merely intelligent who know many facts and can do mental manipulations but who lack deep understanding and vision. People today want their kids to grow up "smart," but success in life more often goes to the sociable and wise rather than to those who are merely intelligent. Wisdom comes from experience, which means failures, so if you must ever choose between hiring those who are academically brilliant or those who are worldly-experienced then choose the experienced.

Incidentally, it struck me the famous marketing consultant Dan Kennedy searched years for a rule that would help companies hire the best employees and found that when the applicants had parents who owned their own businesses the applicants would have seen their parents do everything themselves without complaint, and thus picked up the trait of willingness to take on many responsibilities and do what needed to be done without being asked. These turned out to be the best employees. Hence, one of his rules for hiring was to ask applicants whether their parents had owned their own business.

The third requirement is the tendency to think for yourself and rely upon your own independent judgment rather than just believing group think, what the public commonly believes, and what you are told by others. Leaders must not easily trust information they receive but must learn to carefully assess it using discrimination to judge the probability of it being true or false. A nationally famous top nutritionist once told me he was skeptical of the claims made about every new supplement that came his way but he forced himself to remain open enough to always test those claims, and often found great winners due to forcing himself to stay open-minded.

As general rules of wisdom you should always test new allies and newcomers before trusting them completely, and be wary of fast talkers who are talking so fast that they are hard to follow. Michael Senoff posted an interview with a "Mr. X" Silicon Valley business success (on his HardtoFindSeminars.com site) who said that he never hired ex-policemen, ex-FBI, and ex-intelligence operation professionals because they had "limited views of positive business ethics." You will often find ethical problems prominent in prosecuting attorneys and lawyers because they are actually trained to lose their moral center, and often marketers or salesman as well because some become willing to say anything to make a sale. Hence you should not just accept the claims made by others, especially in the sales and marketing field, but be remain open enough to still test them unless you know they are blatantly fraudulent.

When you too readily accept external advice this habit will eventually undermine your ability to make sound judgments. A leader must learn to trust his own mind, which might require frequent rejuvenating periods of solitary retreat for rest and private reflection.

As another general rule, a leader should also have an open ear for advice as well as criticism from others. He must remain open to the existence of unpleasant truths. We look at mirrors to straighten our clothes and study history to learn about the rise and fall of nations or whether something has been tried before, but it is through another's honest opinion (or introspection) that we perceive our inherent faults and weaknesses. If we close our ears to criticism or refuse other people's good advice it is far too easy to become ignorant and ill-informed.

Political leaders must learn how to gracefully accept information they don't want to hear as well as helpful criticism and suggestions. They must surround themselves with advisors who can offer worthy counsel but must fully practice independent discernment. While you should consult experts for advice you must also rely on your own independent judgment, which you hone from experience and critical thinking.

In short, blindly following others is not a habit of great leaders. As Master Guan mentions there is no stronger defense and no mightier fortress than that of independent judgment because, frankly, there are many people are usually trying to deceive a political or military leader and he commonly encounters errant information all the time.

When a leader fails to heed the corrective advice of others and becomes haughty with hubris then the heights of power might cause him to become subject to what psychologists call the "dark triad." This is an ego trap of (1) entitled self-importance, (2) a tendency to start strategically exploiting people using deceit and manipulation, and (3) a tendency to exhibit the psychopathy of cynicism and callousness. Some become megalomaniacs with an enhanced sense of self and delusions of grandeur.

Once you rise to a supreme position of political leadership it is easy to fall into self-exaltation and begin to mistakenly think that you are inherently greater than others, more deserving than others, entitled while others are not, and smarter than everyone else ... all because you now seemingly rank above everyone due to a *temporary* artificial status. These failings, and overconfidence, have been the undoing of many.

It is indeed possible to become like a deranged Roman Emperor after you achieve power. I have personally seen many rich people start treating regular people like shit, and have seen people who became rich by luck deceive themselves by attributing their good fortune to having superior acumen.[24] Those who become politically powerful also sometimes start cheating themselves into thinking they are now invincible or entitled to be free of moral restraints, and then they try to extend themselves too far or engage in heinous acts of hubris.

This happens to states, too, for Edward Gibbons wrote that when civilizations reach their zenith with power and influence they start to think they are the noblest and most advanced civilization known in history, and develop a blind sense of invincibility. Great Britain's high status before WWI and WWII, for instance, contributed to a state of arrogance that obstructed its need for progress and growth.

An individual aspiring to political power should always stay balanced, ethical, examine the words of others carefully, keep his eyes and ears open, always ask questions, and *always keep a check on himself,* which you do by taking other people's advice and by practicing introspection. It is only by maintaining a humble open ear that a leader can correct himself and remain close to the people by discerning their concerns.

The wisest of leaders focuses his concern on things not yet manifest, and trains in being "perceptive" by noticing subtle phenomena and divining what they foretell.[25] He acts at the first sign of worrisome symptoms to

---

[24] This popular self-deception, which is related to the fact that power corrupts, is a rather interesting topic in itself. For instance, wealth often misleads people in the same way that political success makes people think they're omnipotent and invincible. No individual wishes to think that his wealth or status is a fortuitous affair; everyone wants to believe that their wealth or standing is the result of their own superior abilities. Hence in attaining some above average status an individual will often start believing in his own press clippings, close his ears to others and then fall. The public also falls subject to its own delusions, such as the general fiction that intelligence, wisdom and perception march in line with the degree of someone's wealth. Since money and consumption are the measures of capitalist achievement, a common view is that the more one's money (representing progressively greater achievements) the more intelligence there is to support it or the deeper the individual's mental processes and perspicuity in sizing up situations and judging future trends.

[25] As an example, during China's ancient Shang dynasty the nation was a struggling agricultural state. Hence the people's tools and utensils were for practical purposes and lacked excessive decoration or ornamentation. One day the wise brother of the emperor noticed that his Majesty had started to use ivory chopsticks when eating his meals. "The dynasty is doomed," he surmised, "for if his Majesty now uses ivory chopsticks for eating soon he'll feel that earthen dishes are not good enough to match them. Next he'll want costly cups and other expensive trappings as well. No one who eats grains and cereals with ivory chopsticks will be content to live in a shabby house so soon everything, from his clothes to his residence, will have to become excessive to match them. Since our country is poor and built on agriculture we cannot afford such luxury and if we waste our money on such

make sure that trouble never arises. Like a circus acrobat on a high-wire, he senses imbalance quickly and acts to compensate for it in time. What others don't notice he does notice, and he acts to restore balance immediately. By emphasizing prevention and immediate correction he avoids a thousand ills. He should try to preempt everything that can go wrong.

"The superior man worries over misfortunes that have not yet become apparent," and so he acts to correct errant conditions that disasters might never strike. He fears the hidden, internal aspect of any situation rather than its external ugliness that has already been revealed. After all, it is impossible to defuse a bomb that has already exploded. For this reason he observes what is in people's hearts because this is the key to their latter behavior.

If you put your energy into arranging situations so that they run smoothly then problems will rarely arise. The goal is to stay at ease through the art of prevention and then run things without a hitch. The stupid man, on the other hand, is forever in reactive mode and seems endlessly caught up within a chaos of troubles. He is always running about trying to fight some blazing fire and is afraid to hear of problems that have already manifested.

The supreme leader avoids this fate because he envisions possible dangers before he ever acts. What others learn at the end of affairs he surmises from the very start. He cautiously plans his affairs as if he was avoiding some final calamity, and because he acts with care and foresight a crisis never sneaks up on him and he never faces the danger of extinction. The foolish, however, blind themselves to the potential of unfortunate events and danger.

The standard method to obtain guidance for the present is to study what has gone before. By reviewing history you can determine what the future likely has in store, what opportunities can be expected, and what troubles have to be headed off from the start. A famous Chinese saying aptly runs, "Do not forget the past because it is the teacher of the future." Thus the wise leader is never fooled with the words, "This time the situation is different," but studies the repetitive patterns of human nature

---

personal luxuries and enjoyments our neighboring countries will become envious and will then attack us. In either case, the people will become either the miserable slaves of an avaricious emperor or of an ambitious enemy. No matter who rules in the end, war, poverty and chaos are around the corner." Less than a year later, the emperor ordered thousands of farmers to build several grand palaces with gardens, and the emperor started to live lavishly while the people suffered miserably. In time a great lord from the West (King Wen) came to challenge the emperor, and by attacking the state he overthrew the dynasty. The entire situation had been surmised from the beginning simply from noting the appearance of a set of ivory chopsticks. Such keen insight and analysis is the mark of genuine wisdom.

inherent in on-going situations that are creating the cycle of world events. He enhances his chances for gain by being always ready to make use of expected opportunities and tries to ride the tide of present events to successfully accomplish his goals.

An example that comes to mind is the case of Count de Marenches, long time head of French Intelligence, who received information at the end of November 1971 that the Americans were going to devalue the dollar. He immediately contacted President Pompidou, who was once a French banker, to inform him of this coming event. Marenches wrote about what happened next: "While from this point (onwards) I was out of the information loop, I was able to follow the outlines of the action through our own monitoring services. Working quietly through a variety of what can only be described as 'cut outs,' or individuals whose ties to the government were deniable, the Banque de France, France's central bank and equivalent of the United States' Federal Reserve, was able to put into effect a series of operations that proved highly successful. By quietly selling dollars and buying francs in a number of markets around the world, the central bank was able to accumulate enormous profits – by themselves, enough to have financed all the operations of the (Intelligence) Service far beyond my tenure in office, perhaps until the end of the century. Were we not profiting from the misfortunes of a friend and ally? Perhaps. But at times, rare though these instances are, that's part of the game."[26] Here the French made use of unexpected opportune timing and covered their stratagem with a shroud of secrecy.

No matter how excellent your planning, however, you can never underestimate the importance of opportune timing for accomplishing great goals. Proper timing lets you accomplish more with less risk, less effort and less trouble. Timing is often of more significant than specific tactics used in government, military and business strategy. Those who are skilled in the art of timing can always accomplish their aims in less time, with less effort and with less difficulty than those unskilled. They pick the most propitious moment to launch campaigns and offenses. Mastering timing is therefore one of the essentials for becoming the greatest leaders.

LESSON 13: The Need for Careful Preparations

In formulating strategies you must rely upon basic principles because strategies devoid of core principles [foresight] will end

---

[26] *The Fourth World War: Diplomacy and Espionage in the Age of Terrorism*, Count de Marenches and David A. Andelman, (William Morrow and Company, New York, 1992), pp. 114-115.

in trouble. Furthermore, undertakings that lack full preparation will come to naught. For these reasons the sage kings always saw to making their preparations complete and would wait for the right chance before acting. Even for a limited aim they would make full preparations and then wait for the right opportunity [to act]. When the proper time arrived they would decisively launch their armed forces and cut through resistance to attack those who refused to yield. Thus they would destroy large states and control their territory. The large states are primary and the smaller secondary. The sage kings assembled the nearby states in order to attack the distant, used the large states to rope in the small, the strong states to direct the weak, and the many [majority] to bring along the few [minority]. The common people were benefited by their righteousness and their fame spread throughout the entire realm. Therefore, the feudal lords carried out their orders uncontested. No neighboring countries opposed them and no distant states failed to obey them.

To accomplish grand objectives you will need a sound foundation of planning and preparation. No one wins the Olympics without extensive prior training and practice, and no one wins wars without a similar degree of preparation. No country becomes rich without a set of strategies either, and no political leader gains power without a set of plans and preparations as well.

World class achievers all know the importance of preparation and practice because prior devotion to groundwork is what enabled them to excel. Without careful preparation, practice and forward thinking there is bound to be failure in grand attempts because great achievements are rarely born from luck. You must plan to achieve them and create a monitoring system to stay on track with corrective perseverance.

The rule for accomplishing your undertakings is to thoroughly master the situation fundamentals, to map out viable strategies, and to then diligently execute the plans you've created with persistence and perseverance. Countless *I Ching* hexagrams stress to political leaders that they need to continuously work at their objectives with strong perseverance and persistence (see *The I Ching Revealed*). The number of times this theme appears in the *I Ching* is astounding.

Lee Kuan Yew said, "I do not want to be remembered as a statesman. First of all, I do not classify myself as a statesman. I put myself down as determined, consistent, persistent. I set out to do something. I keep on chasing it until it succeeds. That is all."[27]

To win or preserve a kingdom a political leader must obtain the best advisors and counselors possible. You can easily put a value on jewelry but it is impossible to put a price tag on those who can provide you with excellent strategy and management that wins you your life objectives. They are more valuable than buckets of gold. Whether you wish to achieve grand objectives or limited aims it is always right to seek out the talented and make solid preparations for your plans.

No amount of planning can guarantee success in your endeavors or eliminate their risks entirely, but careful planning can definitely decrease your chances of failure. Wise advisors know the pitfalls you need to avoid. As in the investment field, the primary focus is to prevent losses as you try to make gains rather than just focus on making profits. Investing is an art of risk management until gains come along while politics is a game of survival until ascendancy can be achieved.

An often quoted maxim is that the right actions at the right time can overcome almost any obstacle; the proper plan of action for the proper place and time can surely produce success. Napoleon attributed half of his military success to the fact that he could motivate his men, and the other half he attributed to the fact that in planning he could calculate how long it would take to transport a herd of elephants from Paris to Cairo. Zhuge Liang was known as a strategist who always calculated his logistics carefully, and was known for always thinking one step ahead and planning for those eventualities. Hence, once again the importance of planning is brought to the forefront.

In war a general relies on the greater to overpower the lesser, the quicker to triumph over the slower, and the many to conquer the few, so a leader must also rely upon the relationships between yin and yang in order to plan his affairs. He views his designs as if they were action plans for fighting battles knowing that no general enters battle without prior preparations to take advantage of the circumstances. He doesn't base his plans on fine sounding theories but on the behavioral tendencies of human nature that repeat throughout history so that he can use that constancy in his own plans and expectations.

In considering the relationships between the great and small, superior and inferior, strength and weakness, reality and appearance, conventional and unconventional and so on, an excellent leader will use the big to lead the small, the strong to destroy the weak, and the many to help the few when trying to benefit the world. He will suppress a powerful country to help a weak one, will stop tyrannical rulers in order to save their people,

---

[27] *Lee Kuan Yew: The Grand Master's Insights on China, the United States, and the World,* Graham Allison and Robert Blackwill, (The MIT Press, Massachusetts, 2013), p. 158.

and will rescue people teetering on the brink of destruction. All these actions further the peace of the land in addition to bolstering a leader's reputation and standing.

When a leader's actions ("virtue") nourish all the people of the world in this way then his fame will spread everywhere and he will be able to successfully lead nations other than his own. This is how to gain political power and influence and cause other countries to follow your lead, which is by not acting in your own self-interest.

The key behind this accomplishment is making careful strategic plans and employing advantageous timing. When you cautiously employ both these factors then you can act without regrets.

LESSON 14: Leadership Actions That Win Universal Approval

> For an enlightened king to naturally become the leader of the world is only reasonable and logical. He represses the powerful states, assists the weak, restrains those rulers who become tyrants and stops those who are corrupt. He revives states that have collapsed, stabilizes those that are endangered, and restores the broken lines of succession [government]. These actions win the support of the entire realm and turn the feudal lords into allies. These actions are also regarded as a great blessing by the common people. That is why the entire realm makes him its king.

Why is it that an individual can become the leader of the world? It's because he considers the people as a single family and people see him unselfishly devoting himself to solving their problems because he is concerned with their welfare. He assists those in trouble, punishes those who mistreat others, teaches people the merits of reducing desires and moderation, and acts to stabilize dangerous conditions.

In short, the story of his virtues spreads everywhere because he takes actions for everybody's welfare. It is normal for personal beneficence and benevolence to be reciprocated with admiration and respect, and when you are consequently trusted and respected by everyone it is natural that you are made their leader. This is the gist of Master Guan's path for gaining power and influence.

What guidelines does this suggest as regards statecraft in general? If a country wishes to be well-regarded then it must treat all countries with courtesy and respect by considering them as equals. Even those countries that are small should be accorded dignity and respect, and treated with generosity, patience and an accommodating spirit.

All things have their time and so a state deemed insignificant today can assume tremendous importance tomorrow, especially in warfare. Therefore, it is always best to keep on good terms with every nation. You need only remember that it is not the most successful individual but the individual who is successful *and on friendly terms with others (well-liked)* who usually receives promotions within corporations.[28]

The lessons of history show that nations wishing to become world leaders should maintain principled relations with other states, no matter their status, because in this way they will be considered respectful, trustworthy, and benevolent. If a great nation maintains this consistency of manner it can hold onto the mantle of world leadership even if it suffers temporary upsets such as severe economic decline.

LESSON 15: Advisors Worthy of a King

As for those whose knowledge spans the entire world, whose works endure throughout the age, and whose talent awes the four seas—these are the ministers who can aid a king. If a state possessing only a thousand chariots obtains the guidance of such ministers, the feudal lords can be subjugated and the entire realm can be subdued. However, when a state of ten thousand chariots does not have such ministers it will lose its power in time.

Exceptionally talented, honest men of ability who have superior general knowledge, display broad vision, can develop strategies beyond the normal ken, are wise about the world, highly skilled in decision-making and cognizant of what has remained supreme throughout the centuries—such are the men qualified to be counselors to a king. These are men of mature wisdom with wide mental perspectives because they know history, have traveled extensively outside of their country, and have held a variety of offices and positions so as to obtain a wealth of experience. Therefore they know what works and what doesn't work and they also know how to observe, assess and transform situations.

<u>When you are about to make</u> an important decision it's best to look at

[28] Here we must note that there are many individuals at the top of large institutions who are not there because of exceptional skills but because they least offended others during their climb up the professional ladder. In other words, they were the least inimical among the contending talent, and hence the most likely to rise because of a lack of criticism and ill will. In a highly bureaucratic organization, the lack of offensiveness, despite mediocre talent, often elicits a higher position in the group.

all sides of the issue and to hear a variety of opinions from such men. The goal is to use their wisdom to establish your own personal success so you surround yourself with men smarter than you are and take heed of their advice. It is a great mistake not to consult advisors whose wisdom, judgment and competence are unsurpassed in the land just as it's a mistake to favor mere flatterers over the prudent and experienced.

Furthermore, you should always consider their advice with your own independent judgment. You must evaluate matters carefully according to the Roman dictum *lux vitae ratio*—logic is the guide of life. Otherwise, you will be easily swayed by the liberal "idealistic thinkers" who advocate beautiful words that please but which are empty of practical truthfulness.

Lee Kuan Yew once said, "My life is not guided by philosophy or theories. I get things done and leave others to extract the principles from my successful solutions. I do not work on a theory. Instead, I ask: what will make this work? If, after a series of solutions, I find that a certain approached worked, then I try to find out what was the principle behind the solution. So Plato, Aristotle, Socrates, I am not guided by them. ... I am interested in what works. ... Presented with the difficulty of a major problem or an assortment of conflicting facts, I review what alternatives I have if my proposed solution does not work. I choose a solution which offers a higher probably of success, but if it fails. I have some other way. Never a dead end."[29]

Unfortunately, idealistic policies based upon unrealistic views of human nature—such as Marxism and Communism[30]—have in the past lead many a nation astray, killed hundreds of millions of people, and will continue to do so in the future. It's not that Socialist-Marxist ideas were untested and deserve a new chance again but that they had been tested

---

[29] *Lee Kuan Yew: The Grand Master's Insights on China, the United States, and the World*, Graham Allison and Robert Blackwill, (The MIT Press, Massachusetts, 2013), p. 135.

[30] Throughout history people have always been attracted to the concept of a utopian society on earth. Unfortunately, this will never exist although philosophies will arise again and again to refresh this seductive view. For instance, nations used to believe that *Christianity* was necessary for social progress and economic modernization. Then they thought that *science* was the solution to these issues. Now the common belief is that *democracy* is the cure-all to prosperity and development. Some nations, such as China, take *materialism and Communism* as their foundation for the creation of a perfect and progressive nation, but all of these concepts are wrong. Perhaps the Japanese are closest to the mark because they recognize that there will always be change in the world and you must employ discipline and hard work to progressively adapt to whatever happens. The idea of a one world government curing ills is wrong as well for there were always civil wars within the single Roman Empire. A one world government would fare no differently.

many times and always fail because they ignore the fundamentals of human behavior. It is not that people should have equal outcomes but *equal opportunities* and then given the chance to excel and rise, which will move society forward. No one ever benefits from them except a small cadre of elites at the top of the control pyramid. Also, those political systems have killed more people than any others in history.

Good advisors help ensure that nations avoid such catastrophes but instead rely on the best that thinking and history have to offer.[31] This is why Confucius said, "If a man is able to review the past and know the new, he may be the teacher of others."

Even though a leader may not like the worthies he employs, he should forget his personal feelings and use those talented for the benefit of the country. After all, the fate of a nation largely depends upon getting the right men, so the importance of talent cannot be emphasized enough. For instance, the Chinese Duke of Zhou was so fearful of missing capable men and useful advice that he would drop the food in his mouth and run to his door to receive any visitors who might come calling. His actions showed that he wouldn't risk losing the chance to employ any man of talent, which is the same attitude that a premier political leader must adopt. They work for you so you must not fear them but use their talents to help yourself succeed.

A medium-sized country making use of capable men will be able, at the very minimum, to maintain its standing and influence in the world. Without talented ministers, however, even a great country will in time lose its power and standing. In the business arena, you can say the same thing about the need for great managers.

Where does one find such ministers, managers and advisors? Since the straightest trees are most often found in the remotest forests they often are to be found in quite obscure places, far from where one would expect them. It was during a hunting expedition in the countryside that King Wen found the sage Lu Shang Taigongwang (Jiang Shang or Shi Shangfu) fishing on a river bank.[32] Liu Bei, founder of the kingdom of Shu, sought out the

---

[31] A typical problem with American advisors is not only their insistence on viewing the world through "American eyes" (a problem of ethno-centricity faced by many countries), but the fact that they tend to collapse politically significant time into only the last twenty years. The most excellent advisors, however, are familiar with cultural and historical patterns of several millennia, which is accordingly reflected in their opinions. Nations have memories long enough that they bear grudges past the two centuries mark, and wise statesmen understand this.

[32] Incidentally, Lu Shang felt that a country could become powerful only when the people prospered. He felt that the ruler would not last long if the officials became enriched at their expense and the people remained poor. In his eyes to "love the

reclusive political-military genius Zhuge Liang in the remote countryside as well. Guan Zhong was found working for Duke Huan's rival.

While at times you can use the method of worshipping the skull[33] to entice talent to come to you, at other times you must humble yourself and set out to find the real talent of the nation. These men are sometimes hidden and may even appear quite different in appearance or behavior than the self-seeking who usually surround a nation's capital since their thinking is independent and they often do not conform in various ways.

Therefore a man seeking political power must search for exceptional advisors so that he can make use of their collective wisdom. If he's wise enough to go out of his way in this manner he will also be wise enough to ignore the remote theories and fine sounding words he's offered and prefer sound common sense by such men. He will wisely base his policies on the realities and fundamentals of a situation—what is simple and basic and functions throughout the generations. He will craft his policies to address the fundamental causes operating in a situation.

Individuals fashioning policy must recognize that subsidizing behavior will produce more of that behavior, penalizing behavior will cause less of it, attempting to eliminate one problem unskillfully will often cause other unintended problems to arise, and that one's strategies must adapt to new circumstances. A cardinal rule is that incentives provide motivation while punishments produce deterrence, and the most effective reaction to opponents who harm you is the strategy of tit for tat (equivalent retaliation).

The greatest plans succeed or fail because people remember or neglect simple principles such as this, so a leader must take them into account when fashioning his policies and plans. The fundamentals you need to rely

---

people" meant reducing taxes and conscripted labor, so by employing these principles King Wen made the Zhou state rich and powerful very rapidly.

[33] Once King Zhuang of the Chinese state of Chu was visiting the Marquess of Shen but wouldn't eat the meal that had been set. The Marquess, noting a worried look on the king's face, feared he had done something wrong and asked for the king's pardon. King Zhuang only replied, "You haven't done anything wrong. I've heard that a capable ruler with good advisors can be a king, and a ruler with medium talent but able advisors can maintain his influence. But if a ruler is of mediocre talent and has inferior advisors, I'm told he will lose the state. I'm of low ability myself and lack able men to assist me, so I worry about losing the country. While there are many able men about, I'm unable to enlist them, so what's the use of eating?" When the news spread that the king was so desirous of enlisting worthy men that he couldn't eat, a variety of talented men stepped forward offering their services while many other distinguished individuals were recommended to the state. This is another example of the method of "worshipping the skull."

upon should be simple basic principles that underlie human behavior.

While one might not always be able to find a Guan Tzu, Jiang Ziya, Zhuge Liang or Bismarck for advice[34] there are always talented worthies who are capable of helping to lead a state. Super talented people are a great blessing but in many cases you simply need competent people. Peter Drucker once said, "No institution can possibly survive if it needs geniuses or supermen to manage it. It must be organized in such a way as to be able to get along under a leadership of perfectly normal human beings." The billionaire investor Warren Buffett said, "I always invest in companies an idiot could run because one day one will." You don't need to employ sages in office because as long as your administrators have intelligence and capability then your country (or organization) will run well.

Virtue is an important factor in this equation. Castiglione once warned that when goodness is lacking then sagacity is mere cunning so the individuals you employ should have good characters. You are looking to employ honest men who treat people fairly, are loyal, just, prevent harm, and of course are intelligent, skillful and capable. Being talented includes possessing the virtues of character.

When a ruler can find capable men of good character and employ them in office then his country, no matter how small, will not fail to become strong. Nations which do not employ gifted officials, however, will see their power and standing slowly erode away.

The same can be said of any business as it is the people running a business who will make it a success or failure. Bad managers have fouled up plenty of great businesses, and wonderful managers have transformed many a lame business into a good one and good companies into world class organizations.

Never doubt that a few strong men can bring about tremendous change. Cortez and but forty men toppled the Aztec empire. It was just a handful of men who brought about the personal computer revolution. A small group of men risked their lives and created the United States Constitution. Margaret Mead reminded us that a small group of people can indeed change the world. A leader should therefore surround himself with able advisors and capable men, and all his important undertakings should begin with their counsel.

LESSON 16: Certain Administrative Deficiencies Will Lead to Ruin

If all other states in the realm are well-managed while your own

---

[34] Or an enlightened Joseph who could inform the pharaoh that seven lean years would follow seven fat years of plenty.

state is in turmoil and upheaval then you will soon lose your state. If the feudal lords [of the other states] are all on good terms with one another while your own state stands apart in isolation then you will soon lose your state. When neighboring states are all well-prepared for war but yours remains at ease then your state will soon cease to exist. These three things are the signs of a state that will perish.

To gain insight into the future of your country you need only compare it with its neighbors and those of similar circumstances. The Atlas of Economic Complexity (atlas.cid.harvard.edu), for instance, can help you judge your economic robustness beyond the normal routine of simply comparing GDPs. If your state ranks poorly compared to its peers in various ways, especially in economics, then it needs to think carefully about its policies.

When an organization stands isolated from its peers in terms of the way it does things, and suffers adversity while its colleagues enjoy prosperity, it is broadcasting to everyone that it will soon cease to exist. History shows a large string of countries that collapsed due to adopting the wrong policies, which even happens in modern times![35]

The economic, military, political and other conditions under which countries have collapsed have varied widely throughout history. Each case in the past was usually subject to multiple factors such as soil degradation, deforestation, tribal migrations, foreign invasions, changes in technology, changes in the methods of warfare, changes in trade routes, depletion of mineral resources, excessive debts, excessive social welfare spending, cultural decline and social decadence, popular uprising and civil war.

An interesting experiment in the 1960s by John Calhoun should be noted. It was once repeatedly run on both rats and mice where he produced a "Mouse Utopia" or giant welfare state where the rodents were provided with all the food and water they needed, ample living space and an absence of predators to see what would happen to the population. This was like providing people with a utopia of cradle-to-grave social welfare and has implications for countries that might want to head this way.

At first the mice did well, and their population skyrocketed due to the unlimited resources. However, eventually there came a turning point in

---

[35] The Chinese classic, *The Romance of the Three Kingdoms*, summarizes this nicely in its opening sentence: "Empires wax and wane; states cleave asunder and coalesce." In the last 500 years, out of dozens of empires that have flourished there are only four states—Portugal, Holland, the U.S. and Britain—which have been durable enough to sustain their present form for at least 100 years. There is nothing to suggest that democracy is the cause behind this longevity.

mouse utopia. The experiment was repeated several times but a breakdown always eventually occurred in social norms and structure that caused a total collapse of the entire population and then complete extinction. At a certain point the male mice started becoming feminized and no longer defended their territory. Female mice started abandoning their young, became aggressive and started assuming the roles of males. Violence became increasingly common and the social disorder skyrocketed. Male men began to assume female roles (transgenderism) and male homosexuality emerged while pedophilia (where the male mice mounted the young) grew rampant. Female fertility decreased and mothers started rejecting their young, or showed no interest in sex. The mice started avoiding all stressful activities, including competition, and focused their attention increasingly on themselves. Non-reproducing females became preoccupied with eating, grooming and sleeping – basically maintaining the physical attractiveness of a healthy, well-kept body. They had wonderful outer appearances but lacked cognitive and social skills. They looked inquisitive but were, in fact, very stupid. They were unable to reproduce, raise young or compete for anything. They self-isolated or over-indulged themselves and lost interest in anything that perpetuated the species.

A utopia where you have everything available at any moment for no expenditure or risk basically leads to social abnormalities and then population extinction. People need challenges and need to pass on valuable skills to the next generation of young in order to thrive.

Now imagine if the powers that be within a country, say a United States, started to use social media to create within the general public a wider adoption of *the very same characteristics that produced population decline in the mice utopia* such as the WOKE and LGBT issues concerning male feminization, homosexuality, pedophilia, transgenderism, female masculinity, etcetera. Let's also imagine that the government made moves to weaken religious and racial cohesion among the people, the involvement of parents in their children's educations, and made moves that weakened the fabric of strong families. Let's further imagine that for various reasons the country began encouraging strong trends in the younger generations of self-absorption with one's physical appearance, a decrease in socialization skills, an avoidance of competition (in order to keep self-esteem high), declining responsibility for personal behavior, and so forth.

If these aberrations were actually being pushed on a population through control of the media narrative (including censorship) – the *exact same very ones* that cause a mouse utopia population to terminate – it might seem as if its leaders were either extremely stupid or actually trying to destroy the country's population on purpose, wouldn't it?

It is strange that Russian President Vladimir Putin noticed this but our own leaders do not. During a speech on Russia's annexation of Ukraine he addressed his nation and asked, "Do we really want to see perversions that lead to degradation and extinction be imposed on children in our schools from the earliest years, for it to be drilled into them that there are supposedly some genders besides women and men, and offered the chance to undergo sex-change operations?" He said that western elites were pushing "a radical denial of moral norms, religion and family" that was "a complete denial of humanity, the overthrow of faith and traditional values."

There are many causes behind states that collapse, but could a nation pushing these things on to its own population actually be willingly engaging in self-termination?

In terms of internal stability, economic development, foreign relations and military preparedness, no state can long lag behind its peers without facing the possibility of extinction. Why bother to pursue excellence in these areas if you are interested in self-extinction?

Nations also need allies for their times of trouble, and must possess a strong military to dissuade invasion by others. They must always be prepared for military offense and defense, namely the possibility of war. War, pestilence (plague) and starvation are the three main factors that have destroyed states in the past.

LESSON 17: Trends at Extremes Always Reverse Their Direction Where Yin Turns into Yang and Yang Turns into Yin

If a state is large but its government is unimpressive it will eventually become smaller and smaller. However, if a state is small but its government is impressive it will gradually grow larger and larger. If a state grows large but shows little achievement it will become smaller again. If a state grows strong but is not well-governed it will become weaker again. If a state develops a large population which cannot be employed then its population will eventually decline. If a state becomes prestigious but neglects propriety it will lose the respect it has earned. If a state becomes important but acts intemperately then it will lose the influence it has gained. If a state becomes rich but turns arrogant and dissolute then it will again become impoverished.

When a large state accomplishes little it will end up being criticized and slandered by others and will eventually diminish in size as was seen with the collapse of the U.S.S.R. On the other hand, if a smaller state accomplishes a great deal, as was the case with Singapore, then it will gradually develop and gain prestige accordingly.

Size is not a requirement for gaining power and influence because a small state showing exceptional accomplishment, such as a Singapore or Japan, will be accorded due recognition for the wise management policies of its leaders. If a small country can cultivate prosperity and wide vision it can eventually become a powerful nation. But if a large and powerful country elects to follow mediocre policies then it will end up falling into that mediocrity. It is all a function of its policies.

One of the natural laws of the world is that all phenomena are fated to change, meaning that a turning in the tide of present affairs is at some time bound to happen. Situations will always change into their opposite wherein a Disney that traditionally promoted family values will start promoting the opposite. Despite all our efforts to institute laws to prevent instabilities we must accept the fact that the world is constantly changing and that there is no natural security in the world for even a single moment.

All worldly conditions are marked by impermanence and will tend to oscillate between extremes.[36] When the sun reaches its zenith it will begin its decline; when the moon becomes full it will begin to wane. The natural order is that conditions reaching an extreme will reverse their trend and start heading in the opposite direction. Economists call this wobbling back and forth a "regression to the mean," but the principle of bobbing between yin and yang involves more subtleties than that.

Just as we say that "pride goeth before a fall," when something becomes too full this means it will soon dissipate. A sport star's fame might reach a peak with all sorts of positive news coverage and even a *Sports Illustrated* cover, for instance, that then marks the highpoint of their career. A city might build the largest skyscraper in the world just before that nation plummets into a multi-year recession, or a corporation might pay to name a sports stadium after itself right before it goes bankrupt. The field of contrarian investing notes all sorts of cases where stock valuation extremes marked the beginning of great declines and thus the peak of yang leads to yin.

If something rises too high that means it must fall. Thus stocks will rise and fall in alternation, busts will be followed by booms and booms followed by busts, and *recessions (or depressions) will always reoccur because you cannot get rid of the business cycle* despite the best efforts of governments. Economies, like drunkards, are fated to move in wavy lines and this also applies to mass movements in human behavior.

It is therefore in the best interests of a country to enact policies that prevent conditions from quickly growing to an extreme because the excess

---

[36] An extreme to be seen in Capitalism is that it produces a wealthy privileged class that turns about and rejects the source of its own wealth to pursue a socialist order, which always fails. When the timing is appropriate every movement or trend produces a reaction in the opposite direction.

of any condition is a state of dangerous imbalance. The danger is that when conditions reach their extremes they often reverse through destructive collapse. For instance, the experience of centuries (*Manias, Panics and Crashes* by Charles Kindleberger) shows that speculative financial bubbles tend to arise over time and usually end with a pop that produces financial loss and destruction.

If a government encourages its people to neglect hard work, prudence and personal responsibility in favor of government handouts and assurances this will actually damage the national fabric of the country. A nation should always encourage people to take responsibility for their own actions and work for their own prosperity. National policies should encourage the people to abandon the idea that others "owe them," and should instead champion the principles of entrepreneurship, prudent risk-taking, self-reliance and caveat emptor.[37]

The wisest government policies are therefore those that encourage public prudence and moderation; they don't concentrate on trying to eliminate all public risks but encourage people to become personally responsible for their own risk-taking. It might seem an admirable goal but a nation which tries through legislation to achieve absolute risklessness for its citizens creates a mommy state of weak people, and is as foolish as a scientist chasing after perpetual motion.

Wise statesmen count on people to pursue their own best interests and adjust national policies as conditions and customs change just as medicine is prescribed according to patient symptoms.[38] Furthermore, just as good doctors do not continue administering medicine when a patient is no longer sick, so too is there no need to continue government programs when they have fulfilled their function.

The best administrative policies fit like a glove in matching the conditions they address. They mimic the adaptive characteristics of water, which assumes the shape of any container that it fills. Just as a river does

---

[37] If you examine the history of financial speculation you will find that no amount of regulation will protect an individual's mind during periods of euphoria. It's only through the possession of common sense and a "wise head," in contradiction to and defiance of those about you, that you can escape the mass insanity that grips individuals, financial institutions and the government alike. In fact, during euphoric periods of fantastic valuations the group think is extremely protected and sustained in order to justify the forces artificially enriching everyone. Were you to interject doubts about this trend or protest public headiness then as the bearer of bad news you would be ostracized, ridiculed and condemned by these groups. No one ever applauds the man who announces a coming bear market.

[38] The key point is to first inquire into causes and conditions before administering treatment. When conditions change, the treatment should be altered as well.

not run straight forever, good administrators are also flexible enough to change national policies with the times. Foolish people may not understand the wisdom of reversing policy at appropriate junctures but you must adapt your policies to change with the times. A perfect example is the fact that a nation's birth rate changes over the decades.

All worldly conditions eventually change and when they do it is time for government officials to abandon what is no longer appropriate and alter policies to take account of the new realities. Singapore is a state that successfully follows this dictum for it is ready to reverse government policies when it finds they are not working or no longer necessary.

As another example, the founders of the U.S. Constitution acknowledged that America's form of government was a "grand experiment" that might not survive and might require future alterations, which is why we have a Bill of Rights and Constitutional amendments. They knew this because democracies, by their very nature, encourage internal dissent that can tear their homelands apart. Therefore they laid out the means by which amendments might be made to its constitution.

The American forefathers acknowledged that the structure of the government could change with the times and made arrangements for this to happen. One must be especially careful in insuring the social stability of democratic nations because their own inherent right to dissent carries within it the potential seeds of their own destruction. No empire or political system lasts forever—Heaven has never granted the boon of perpetuity to any dynasty or kingdom—and so a nation must indeed be careful of the policies it promotes. Some policies will help a nation grow and prosper, and others will hasten its decline.

Because forces which grow to an extreme will eventually transform into their opposite, a powerful state will weaken as it abandons the discipline that brought it to greatness. *Culture, Country, City, Company, Person, Purpose, Passion, World* reveals how nations lose their founding impetus over time due to laxity and luxury in subsequent generations beyond the founders. Wise states enact legislation which encourages moderate or conservative behavior, and they try to head off problems before they even start because it is always easier to prevent problems from happening than it is to fix them. A dollar spent on prevention is much cheaper than any money spent for repair or relief. A great leader should try to preempt anything that can go wrong.

A rich country, as its population turns lazy and over-indulgent due to its accumulated wealth, will eventually see its riches being drained away as the public abandons the hard work that brought it riches in the first place. A state grown too large (in population) will lose its people due to some eventual famine or strife that its increased size engenders. Therefore, Master Guan says that when a nation does not manage these several factors, which contribute to national order, it is headed toward destruction.

In short, any condition reaching an extreme will give birth to a trend in the opposite direction. In Taoist terms, the extreme of yang will give birth to yin and the extreme of yin will give birth to yang. Thus it can be understood why a labor union that pushes too hard to advance its members' interests can eventually become so costly in terms of salary increases and protective work regulations that it ends up destroying the very company or industry on which it depends. In the desire to maximize self-interest it can actually end up destroying itself.

The goal of wise policy is to therefore first bring any organization to a state of stability and prosperity, where yin and yang are balanced, point it in the right direction of developing towards progress and then address any factors threatening the deterioration of prosperity as soon as they might appear. The goal is to maintain a state of auspiciousness for as long as possible before the onset of inevitable decay.

Only nation-states which prevent a culmination of dangerous extremes—through cultivating balance, peace and equilibrium—will be able to prolong their prosperity and delay inevitable decline. To understand these matters, strategists and statesmen must learn to think in terms of yin and yang because these basic principles explain the tides of affairs and indicate the policies one should use to adjust a situation.

LESSON 18: Leaders Should First Observe and Afterwards Assess

If you wish to know what a state is like then observe its sovereign. To know the condition of its army observe its generals. To determine a nation's state of preparation for war observe its farmland [productive resources]. If a state's prince appears wise but is not; if its generals appear talented but are not; if its people appear to be farming but are not—if these three essentials for preserving a state are missing then such a state will not last long.

To be able to determine the future outcome of any situation you must start by gaining knowledge of the current conditions. A doctor, for instance, seeks to know his patient's condition by making inquiries about his symptoms. Only after a proper diagnosis is it possible to determine the disease and its progression as well as the appropriate treatment.

As practitioners of the Chinese I *Ching*[39] know, to accurately predict

---

[39] The I *Ching* is one of the four basic books of Chinese culture. The principles embedded in the I *Ching* can be used for a variety of purposes such as moral

the future you must start by first comprehending the current situation. From knowing where you presently are within a general sequence of affairs you can then know where you are probably going based on likely trends. Observation, or awareness, is thus the first step to contemplation. As Jiang Ziya once advised King Wen when evaluating his political opponent,

> You must look at the Shang King's yang aspects, and moreover his yin side, for only then will you know his mind. You must look at his external activities, and moreover his internal ones, for only then will you know his thoughts. You must observe those distant from him, and also observe those close to him, for only then will you know his emotions. ... [In Shang] the people muddle and confuse each other. Mixed up and extravagant, their love of pleasure and sex is endless. This is a sign of a doomed state. I have observed their fields—weeds and grass overwhelm the crops. I have observed their people—the perverse and crooked overcome the straight and upright. I have observed their officials—they are violent, perverse, inhumane, and evil. They overthrow the laws and make a chaos of the punishments. Neither the upper nor lower ranks have awakened to this state of affairs. It is time for their state to perish.[40]

Therefore, to evaluate a nation and forecast the outcome of its activities you must visit its cities and observe its people. To determine its military strength and readiness you must observe its armies and then analyze its capabilities. To evaluate its productive capabilities you must examine its industry and workers. These conditions are the compass of business, social, military and political affairs.

How many goods (GNP, trade balance, capacity utilization, etc.) does the nation have available? What are the nation's monetary and financial conditions? Is the nation self-sufficient in industry and agriculture? How large and how well-educated is the nation's population? What are the people's habits and desires ... what do they like and dislike? What can you say about the leaders of the people and the advisors to the government?

When you examine the roots of a tree you can predict whether the leaves and branches will be healthy. In a similar fashion, if you therefore discern leaders who appear wise but are not, generals who seem capable but are not, and people who seem skillful but are not, then you have

---

cultivation, scientific understanding, and even divination.

[40] *The Six Secret Teachings on the Way of Strategy*, trans. by Ralph Sawyer, (Boston: Shambhala, 1996), pp. 48, 82.

identified a country that is in danger.[41]

Just as investment analysts scrutinize a company's products, management and financials in order to predict the future of its share prices so one can observe the conditions of a country to determine how well it will fare.[42]

As I pointed out in *High Yield Investments, Hard Assets and Asset protection Strategies* (chapter 4), you should monitor the following indicators to get a sense of a nation's economic welfare: freight car loadings and truck traffic data (which constitute Warren Buffet's "desert island" indicator of the one number he'd want if stranded on an island), and the velocity of money. The Baltic Dry Index or HARPEX container freight index can help you judge the state of the world economy. Shadowstats.com is the only site offering dependable U.S. economic data like GDP and the inflation rate, and the Chapwood index gives an independent estimate of the inflation rate that you can compare with the one given by Shadowstats.com.

LESSON 19: Excessive Assets Turn into Liabilities

---

[41] Observing a situation is very important, but correctly analyzing what one observes is even more important, for this is the mark of wisdom. In other words, you have to correctly understand the implications of what you see on the surface. An example from the newly discovered Mawangdui *I Ching* texts (*I-Ching*, trans. by Edward Shaughnessy, New York: Ballantine Books, 1996, p. 271) shall suffice to make this clear: "King Zhuang of Jing (i.e., Chu) wanted to attack Chen, and sent Shen Yinshu to go and observe it. Shen Yinshu returned and went to report, saying: 'Their city walls are well-maintained, their granaries are full, their sires are fond of studying, and their wives weave with alacrity.' The lord [said]: 'If it is like this, then Chen cannot be attacked. If the city walls are well-maintained, then their defense will be stout; if the granaries are full, then the people will have enough to eat; if the sires are fond of studying, they will certainly respect their superiors; … and if the wives weave with [alacrity], their resources will be sufficient. Like this, Chen cannot be attacked.' Shen Yinshu said: 'Though it is acceptable to put it the way the lord has, there is also a different analysis from yours. [If] the city walls are well-maintained, the peoples strength will be sapped; [if] the granaries are full, … [text missing] … men; if the sires are fond of studying, they will have external ambitions; and if the wives weave with alacrity, the sires will be clothed but without enough to eat. Therefore I say that Chen can be attacked.' Thereupon they raised arms and attacked Chen, conquering them."

[42] An excellent example of insightful analyses of national fortunes can be found in *Investment Biker* by Jim Rogers (New York: Random House, 1994). This book is highly recommended to anyone interested in understanding how to evaluate a nation's economic prospects.

If a state's territory is extensive but not well-managed this is known as an excess of land. If its population is large but not well-employed this is known as an excess of people. If its military forces are powerful but out of line with what is necessary this is known as an excess of the military. If these three excesses are not curbed then a state will soon cease to exist.

If a ruler's territory is extensive but remains uncultivated then it is not [to be considered] his territory. If his senior ministers are honored but not subordinate then they are not his ministers. If his state's population is large but not unified then they are not his people.

Master Guan warns that when an asset grows too large this can be a detriment rather than a blessing if it is unproductive. A large population without employment, for instance, is actually a burden rather than a blessing. A government must then create policies to pacify an idle populace so it does not transform into a rabble, and in today's world it must spend money to support the unemployed.

Whenever an asset grows to an extreme without being managed well it can easily turn into a liability. In *Rich Dad, Poor Dad* Robert Kiyosaki takes the position that an asset is not an asset if does not put money in your pocket. Liabilities take money out of your pocket whereas assets provide you with a return. Hence if you own a lot of property you are not renting but must make mortgage, maintenance and other payments those houses are actually liabilities rather than assets. The Taoist viewpoint is also that an asset is only valuable if it is actively productive rather than potentially productive, and in the same way money only has value when it circulates. An asset's value comes from its ability to put money in your pocket.

In the strictest sense, what we think of as an "asset" is valuable when it is functioning productively in operational mode rather than simply sitting in an idle state of potential. The Taoist principle is that an item's real time functioning is what gives it an actual value.[43]

To prove this point we need only think of a famine when there is little food available. Some people would rather barter for goods rather than

---

[43] Consider an uncollected debt (accounts receivable) or outdated inventory that sits unsold. Can you truly consider these items assets if you aren't certain that the debt can be collected or that the inventory can be sold? These items may have to be written off so an accountant tries to be realistic when computing their value for his company. National leaders should look at national assets in the same conservative way.

exchange food for money in such a situation, so the value of currency is nothing in such as case. Master Guan states that uncultivated land cannot be considered productive territory and ministers that don't listen to their leader cannot be considered subordinates either because these are assets in name only and not in actual functional fact. President Donald Trump put many officials into office who secretly worked against him and stabbed him in the back, so while he considered them his ministers they were not that in fact.

Master Guan therefore cautions a ruler from thinking his country is richer than it really is because untapped resources count for little unless they can be put into play. According to this principle a country with powerful weapons should be concerned whether or not they will dependably work on the battlefield without breaking,[44] and whether it has sufficient munitions to use them.

Using this principle, an agricultural state that is calculating its productive resources should not count untilled soil as an asset until this land is transformed into working farms. In fact, the land may actually constitute a burden since it may cost money to protect it while it provides nothing in return. Master Guan says that uncultivated land, inactive factories and unemployed workers should not be counted when measuring a nation's strength as the names "land," "workers" and "factories" do not reflect the true state of affairs.

Only the sages are not deceived by names. They measure national strength only by what currently functions in a country because that's the only thing ultimately existing. Of course we must still value those idle things that can quickly be turned to use, but they aren't actual until they are productive. Just as individuals only prove they are friends in times of need the real value of an item only materializes when it becomes actively engaged in productive activity.

This overall teaching offers excellent advice for governments. A nation should not believe itself rich if its assets are not being employed, especially if its population is left idle. A huge population may be deemed desirable but a state will be encumbered by too many people if they are left without productive employment. This, in Master Guan's view, should be considered overpopulation, which we might state was India's problem when it became free of British rule in 1947 and then set out to educate the

---

[44] If you have weapons that easily break down (such as fighter jets) without sufficient replacement parts, or you lack trained operating personnel, can they really be considered weapons? If you cannot depend upon them or they are outmatched then they are useless and hence weapons in name only. This is an important lesson for governments that increasingly turn toward high tech military solutions. As another example, a weapon so powerful that it cannot be used is a deterrent but not necessarily a weapon.

populace, bring it out of poverty, and increase national life expectancy. In *An Era of Darkness*, India Congress leader Shashi Tharoor said that India was one of the richest countries in the world before the British showed up. Poverty existed in India before the British, but in general Indian society took care of everyone. From two centuries of exploitation and looting, however, the British took poverty from a small problem to a national crisis and made India one of the poorest countries in the world when at one time it represented 23% of world GDP.

When a country's citizenry are all productively employed with adequate incomes, however, the country is usually generating revenue and becoming prosperous and wealthy. Creating conditions for business that employ a nation's people is crucial for a country's well-being, which is why wise governments should strive to attract investment (which produces jobs) by making the business environment as favorable as possible.

Government funds originate almost entirely from taxes that have their root source in business and commerce. When the "yin" restrictions on commerce reach extremes—which is a characteristic of the state of California—this only ends up reducing overall business activity that migrates elsewhere, this subsequently decreases government funds, and it then endangers the livelihood of the population they are meant to help. An excess of restrictive business regulations will cause businesses to leave for elsewhere, thus destroying state revenues and employment, because money will always go to where it can find the highest and easiest returns.

Legislators must put aside idealistic socialist notions to realize that water must circulate freely or it will stagnate, and the strong circulatory currents that make up prosperity within a nation will dry up if business is restricted in too many ways.

In today's lightning fast world of international competition an enlightened government is one which makes sure that business regulations—despite ever increasing consumer and labor requests—do not reach extremes that then jeopardize the country's commercial base. Singapore and Hong Kong are states we can study to learn how to legislate well. Sometimes when lousy governments collapse things even work much better because restrictions are removed and it is like a breath of fresh air.

To correctly understand the economy it is favorable to adopt the Taoist notions of yin and yang to note which business conditions must be amplified and which tempered in order to keep commerce flowing. Too many yin restrictions will kill the yang of an economy. Any country that overly institutes "yin" restrictions is also likely to see its export economy whittled away by foreign competition. Another danger is that the resulting increases in unemployment, social disorder and crime may far outweigh any benefits expected from the legalities so instituted.

As the post-WWII histories of both India and Australia have shown, the excessive business regulations they imposed for a while suppressed economic prosperity even though they were intended to "promote prosperity" and they can actually produce economic stagnation.[45] When dismantled, however, the result is a virtual Spring time of economic activity. President Donald Trump demonstrated this when he removed many of the restrictions on the American oil industry and made the nation self-sufficient *and* an exporter of oil production in less than two years.

In light of today's international trade and competition, and the fact that capital can be instantaneously transferred across international borders, countries must be especially careful when enacting their business laws and regulations. Foreign competitors will always be ready to take advantage of poor solutions which nations impose on themselves.

This is why a country needs wise legislators who can recognize that the push for more and more must eventually have its limits. As in the field of personal conduct an individual must know the limits of proper behavior; he must know where to stop. Despite what many will say, some things are best left untouched by legislation for well-meaning laws may easily become a burden by making situations even worse.[46] You do not have to address every issue within a country but should strive to rectify only the important matters and interfere in people's lives as little as possible.

No matter what disgust people may have for the world of business they must remember that business is ultimately the only means for enriching the people and bringing them out of poverty. It is to no one's

---

[45] A similar example can be seen in the case of government spending. When it constitutes only a small fraction of an economy government spending can act as an economic stimulus when increased. When it becomes the bulk of an economy then the economy becomes sluggish and in danger of collapse. When any factor rises to an extreme it starts taking upon itself characteristics opposite to its initial attributes. Why? Because yin and yang are perfect opposites, and the extreme of yang will give birth to yin while the extreme of yin gives birth to yang.

[46] If governments don't restrain themselves they often go too far with legislation. As an example, many years ago several children died in a fire at the home of a baby-sitter who was not watching the children. An outcry was raised and laws were passed requiring a sprinkler system and license for anyone wishing to offer child care services in that city. As time progressed more laws were passed as anything else went wrong. Soon everything was regulated down to the smallest detail. Even the amount one could charge was regulated and fines were imposed for any breach of conditions. Prices were raised to keep up with the regulations. As a result, eventually child care facilities were no longer readily available because they were too expensive for people to use. Then the government had to step in to fill the void it had created and established facilities lacking the personal touch, with poor service and other attendant problems, at the tax payer's expense! When the government starts to regulate an area this is typical of what happens.

benefit to kill the goose that lays the golden eggs through political systems such as Socialism, Communism or Marxism. Governments must indeed institute laws and protections to prevent business abuses and establish fairness but they must also do everything possible to encourage, rather than discourage, healthy business environments that can produce prosperity for a nation's people.

In political terms Master Guan also tells us that when a group of people no longer support their leader they cannot be considered "his people" anymore. A leader cannot exist without followers because without followers you're not a leader. This way of looking at things sheds light on some very tricky situations in the Muslim countries where it's unclear whether politicians or mullahs are the actual leaders of the people. Master Guan says that when the hearts and minds of a nation are torn between two such leaders the people cannot be considered a "single people" (cohesive whole). Without unity among the people there is no harmonious community, and without community there can be no state.

How can a group of people be considered "your group," in a functional sense, if they do not listen to you and follow your orders? If you have ministers who ignore your wishes because they listen to someone else (perhaps themselves) they are not your ministers either.

With these lessons in mind and the experience of history as our teacher, we can see that it's important for a country to become and stay unified, which occurs when national unity is buttressed by a set of common beliefs and customs.[47] It is very hard to hold together a state when it is composed of non-homogenous entities, especially when each group stresses its own individual character. Hence a foremost task confronting the political leader, with the fate of the nation at stake, is to promote national unity and harmony. This topic tackled extensively within *Culture, Country, City, Company, Person, Purpose, Passion, World* and in the teachings of Ibn Khaldun.

In this matter, history shows the great advantage of instituting a single standard language to bind together a realm. For a country such as the United States that is faced with assimilating a tremendous number of multiethnic immigrants, establishing a single national language and educational framework becomes an even more important task for ensuring national unity and long-term survival.

Over the long run, the ideal of democracy alone is simply not enough to bind a nation together. Democracy is not enough in itself to forge a

---

[47] Perhaps originating from a set of shared experiences—such as war, a depression or a common educational system. A further means of instilling unity in a diverse group, often used by skilled politicians, is to postulate some common external enemy. The threat of an enemy will bring about internal national cohesion.

common allegiance. Strife often arises when disparate groups within democracies wish to have their own political state, especially if they are being egged on by outside powers that wish to weaken the country. Because individuals within a democracy are each vested with the right to dissent those nations must especially emphasize commonalities such as language, history, conduct, civic responsibilities and so forth in order to produce a solid union that can survive.

A leader who wishes to be considered "great" by posterity must devote himself to forging national policies with a hundred or even thousand years of benefits in mind. This is more important than calculating their impact before the next election. As we saw with the Roman general Fabius Maximus, it is hard to be a leader who adheres to one's principles when faced with opposition, which will surely arise. As the boxer Mike Tyson said, "Everyone got a plan until they're punched in the face." However, staying with one's principles is the only thing that can qualify one as a true Maximus or "National Protector."

LESSON 20: Imbalances Signify Dangerous Situations

> Anyone who doesn't own any land yet wants to be rich will live in misery. Anyone who is lacking in virtue yet wants to become a ruler will expose himself to peril. Anyone who gives but little, yet asks for much, will find himself isolated. Furthermore, if the executive power of the state is small but there is an excess of subordinate officials with authority, or if the state is small but its capital city is excessively large then the rulership will be in danger of usurpation. However, if the sovereign is revered and ministers are humble, and the senior has high prestige and the respect of subordinates, then orders will be carried out and people will be obedient. This is the highest model of good rule.

In evaluating resources and desires there are two situations a leader must consider. First, individuals who aren't putting potential resources into play cannot realize their value. For instance, if you don't farm fertile land then you cannot count it as land because it is non-productive since idle.

Second, those who possess few resources, while desiring many, will suffer from their unfulfilled desires. It's like looking at a tasty treat in a shop window but you won't have money to buy it. A political leader must learn to control his personal desires and instead should think of the needs of the state. If you are lacking in virtue but want to become a leader then you will be criticized and attacked for every personal weakness you have.

Here Master Guan talks of deficiency rather than the previous lessons on excess. For instance, if you lack initial capital then you will be hampered in trying to become rich because you usually need money to make money. The richest man in the world for his time, Andrew Carnegie, accumulated his first seed capital because he invested $500 in a railway venture that was an opportunity made possible by one of his friends. If he didn't succeed in that first initial venture it is unlikely that he would have been able to go onto other greater business deals due to a lack of starting capital.

If you don't possess sufficient income or assets you might even find it a burden to eke out a living. Basically, if you have big plans and goals yet lack a sufficient foundation to start from you will just end up just making yourself miserable because you cannot fulfill your goals. You won't even be able to take ten steps in the direction you desire.

People always suffer whenever they don't have enough resources to satisfy their desires. They end up in a state of dissatisfaction and discontent. The lack of material things is not the only situation that can cause difficulties because a deficiency in personal virtues can also lead one into jeopardy. For instance, if a ruler lacks the virtue of simple competency in handling government affairs he runs the risk of being controlled or overthrown. If he lacks honesty then he won't be trusted. If he lacks fairness and justice then people won't want to deal with him. On the other hand, when a leader has obedient and trustworthy helpers then he will avoid such dangers.

Cyrus the Great was once faced with a dilemma when he realized that many of his professional soldiers, who commanded large numbers of infantry and cavalry, were more loyal to their closest comrades than to him. He worried about them starting a revolt so he talked to them daily and came up with another idea to win their friendship to make them more devoted to him than to their comrades. The method he used was through the kindness of sharing food and drink with them frequently because this is the means by which the bonds of trust and friendship grow naturally between people. You can strive to create comradeship and community by sharing pleasures.

It is possible for conditions excessive in any way to lead to negative situations because at their extremes conditions always transform to start moving in the opposite direction. In other words, the extreme of yang gives birth to yin, and the extreme of yin will give birth to yang. When yang and yin are balanced, however, then this balance can become a state of auspiciousness.

This is why the American system of government has lasted so long, which is because it is based on an interlocking system of *checks and balances* designed to prevent any part of the government from attaining too much power. The principle of checks and balances discourages any part of the government from dominating the rest to the detriment of the whole.

Master Guan also notes that when a capital city becomes excessively large then this imbalance will also lead to popular unrest (as does an inequality of incomes). Similarly, when there are too many government officials compared to the population then there will also be in trouble because they are all eating the people's incomes. If there are too many lawyers, such as in the U.S., then the nation will naturally be litigious. I once read that the collapse of the Assyrian Empire was helped by an excess of officials and tax collectors so an excess of officials is a sign of trouble. What I know for sure is that the United States military at present has 44 four-start generals to handle an active duty force of less than 1.1 million men whereas it only needed 7 four-star generals to handle 12 million men in WWII, so its military leadership is bloated and top heavy. Is this not a sign of imbalance and error?

History has seen the collapse of countless states when the government started taking too much from the productive population, stopped enforcing laws fairly, and treated the people unjustly with abuse and exploitation. The political elites who rule nations typically tend towards arrogance, vanity, and an undue belief in their superiority and personal greatness over time. This leads to hubris in behavior where the powerful work to enlarge their prerogatives that allow them to oppress and exploit others.

As a general rule, excessive conditions will naturally lead to a deterioration because of the inevitable move towards the opposite direction. When the debt of a nation grows out of proportion to its revenues, for instance, then a financial crisis or bankruptcy (the opposite of the "prosperity" that allowed the nation to borrow) will manifest. For a nation to experience stability and prosperity its leaders must cultivate a sustainable golden mean of moderation and balance wherein all things find their proper places.

LESSON 21: Final Responsibility Always Rests with the Leader

> Should the realm have two sons of Heaven, a single state have two princes, or a single family have two fathers then control would be impossible. When orders are not clear they will not be carried out; if they do not emanate from a single source they will not be obeyed. The people at the time of Emperors Yao and Shun were not born submissive, and the people at the time of the tyrants Jie and Zhou were not born rebellious. Therefore the responsibility for order or disorder in the realm rests with the sovereign.

There should only be one cook in the kitchen, one head of a household, one commander in the army, and one leader of a nation. In terms of leadership matters any number more than one will result in confusion and disorder.

Therefore *a single individual* must always have the final say by holding both final authority and responsibility for the tasks at hand. There must be one individual that the people can look to for ultimate guidance. This person is the supreme leader.

When disorder arises in a nation we have to look for its cause rather than concentrate on its symptoms, and causality will usually be traced back to the level of its leadership, poor policies it creates and bad management. Whether the people are orderly or rebellious is ultimately due to the actions (or non-actions) of the leader(s). Ultimate responsibility rests with the ultimate leader at the top.

There is a story about a time when the Chinese state of Jin once attacked the state of Chu. Chu's king despaired, "Jin did not attack us during the reign of the former king. Since they are attacking us now during my rule, it must be my fault." When the king's officials heard this, they too shamefully responded, "Jin did not attack us during the time of the previous ministers, so it must be the fault of our administration."

The state of Jin eventually heard of this, and said, "The king and his officials are competing to take the blame on themselves, and the king is humbling himself to his subordinates. This means they cannot be attacked!" So the Jin army stopped advancing and returned home in retreat.

Rather than blame each other each group practiced introspection and found where they were somehow at fault. Of course after any faults have been discovered everyone must work to fix matters, and a great leader must always support the impetus for reform. The point of this story, however, is that a supreme leader puts all the blame upon himself while attributing all success to the people.

President Eisenhower, for instance, prepared a brief statement in case Operation Overlord (the Normandy invasion) failed which read, "My decision to attack at this time and place was based on the best information available. The troops, the air and the Navy did all that bravery and devotion to duty could do. If any blame or fault attaches to the attempt it is mine alone."

Whether in business or politics, this is the proper way to lead others. This is why Ike was the *supreme* commander.

LESSON 22: Leaders Must Take People as Their Basic Foundation

A lord protector or king begins by taking people as his

foundation. If its foundation is in good order then a state will be secure whereas a state will be in danger when its base is not in good order. Therefore, if the sovereign is reasonable and wise then subordinate officials will be respectful. If the government rule is fair then the people will be at ease. If the military is well instructed and disciplined then the armed forces will be able to vanquish any enemies. If capable men are employed in office then everything will be well ordered. If the sovereign befriends humane and benevolent associates then his life will never be endangered. If he employs men with official duties according to their worth and capabilities then the feudal lords will respect and obey him.

The administration of a state, or any organization, starts by taking the individual as the basic unit and considers the public benefit as the goal. Since the people are both the basis and purpose of the state the government's mission should include protecting the people, fostering their education, and promoting conditions which will bring them out of poverty and disorder. It is said that only those who serve the people should govern the people. If the people are at peace and all things are in their proper places because of a fair administration of law and order then a state will be in good condition. Without the rule of law and order, however, a country will be weak.

When a leader therefore acts with love and concern on behalf of his nation the people will receive him with obedience and his government will be secure. If he properly employs the officials and resources of his state then the nation will experience peace and order. If the people conduct themselves with moderation—guided by the principles of virtue, propriety and benevolence—then the country will have social stability. If the public does not shy away from hard work and discipline then the country will see economic prosperity.

Basically, good people are the foundation of a strong state. Whenever you're starting a new firm and have the choice of sufficient funds, a good idea, or good people, choose the good people first. People are the basic foundation of any organization because if money and ideas are lacking capable people can still come up with successful ideas and find the money to make things work. A nation should take pains to educate its people to world standards, instill an innovative spirit, nurture excellence and encourage the average to improve. When people are encouraged to give their best then society progresses.

The head of personnel for an airline company once told me that they hired stewardesses in a way that selected for character because they took it as the basis for a good stewardess. They would assemble all the candidates in a conference room and give a speech while recording the faces of the audience. Those girls who would smile throughout the presentation and continue to pay attention were selected over the ones that periodically ignored the speaker, attended to other things or gossiped with others. They tended to also select girls who naturally smiled figuring that it was easier to keep them smiling than to teach non-smilers how to smile. The selection process was all done on the basis of people's character rather than the tests they asked the girls to take.

Lee Kuan Yew said that he selected people for big jobs based on three main factors: whether they had sufficient powers of analysis, whether they had the ability to logically grasp facts, and whether they would concentrate on basic points and be able to extract the main principles from those points and the situation.

In terms of people, if a nation's soldiers are well trained and disciplined then its army will be strong. If a nation's ministers are capable and rightly employed then government affairs will run smoothly. When a ruler is kind and generous then his life will be safe and he will be able to freely use his ministers according to their capabilities with the result that everyone will respect and obey him.

In short, when the foundations of the state are fundamentally sound and everything functions as it should this represents auspiciousness for a nation.

LESSON 23: The Supreme Leader Cultivates Certain Excellences

A lord protector or king owes his existence to certain capabilities. He is superior to others in terms of his virtue and righteousness, wisdom and strategies, his skill in conducting warfare, his knowledge of terrain [or holding territory] and maneuverability [use of initiatives]. Therefore he is able to rule a realm.

No one can gain political power unless he possesses exceptional characteristics, the foremost of which is virtue. Virtue and righteousness, however, must be accompanied by wisdom and intelligence because only clear insight enables you to make the appropriate decisions at crucial moments in time. Wisdom means understanding that in turn gives birth to strategies and solutions. While virtuous behavior can win the respect and trust of people you need wisdom to be able to derive good strategies (as well as recognize proper timing) or know which ones will work and you need power to be able to execute your plans.

If we push this line of reasoning a bit further we must note that stratagems and tactics are superior to pure military might. The famous story of the Trojan horse is enough to prove that wise strategy is superior to brute force. A major focus of Sun Tzu in *The Art of War* is on tricking opponents to make military mistakes, while one of the predominant characteristics of Odysseus in Homer's epic poem *The Odyssey* (about the Trojan War) was using his intelligence to outwit others.

The power of military force will in time triumph over the perils of dangerous geography, and the geographical environment, in turn, is more important (and thus due more consideration) than simple initiatives. This is because an unfriendly terrain can pose obstacles to military advances while good terrain, such as the Eurasian Steppe, makes travel easy for launching military invasions. Advantageous terrain in warfare is sometimes even more important than either strategy or propitious timing.

If an individual holds these five things—virtue, wisdom, military might, geography and initiatives—he will then have the ability to gain power and make his state great. A supreme leader is one who has these advantages, or seeks to develop them in order to gain power. In any event, those who succeed in winning a state must have extraordinary talents such as these and must be able to employ exceptional strategies, including the ability to wage war.

LESSON 24: The Strategies for Gaining Power

He who would excel at rulership relies on the preponderance of large states to suppress others and the power of strong states to weaken others. When strong states are numerous his plan should be to unite with the strong to attack the weak in order to become a lord protector. When strong states are few he should unite with the small to attack the large in order to become a king. When strong states are many it is foolish to talk about becoming their king. When strong states are few it is a defeatist strategy to proclaim the way of a lord protector.

The divine sage king looks at the condition of the realm and knows whether it is a time for action or non-action. He considers the consequences of actions and determines whether they represent the door to disaster or prosperity. [For instance] if strong states are many then whoever takes the initiative will be in danger while those who lag behind will reap the benefits. When strong states are few then whoever takes the initiative will become king while those who lag behind will be doomed. When warring states are numerous then whoever lags behind may become a lord protector. When warring states are few then whoever takes the initiative may become the king.

This lesson is critical to geopolitics.

To successfully manipulate states you must become adept at developing strategies to address circumstances. For instance, if there are many powerful states in contention for supremacy you cannot become their ultimate leader unless the power of those competitors is first diminished. Therefore when strong states are many then becoming their supreme leader isn't likely to happen. The best you can hope to accomplish is to attain a strong influence among the group.

Master Guan says that under these circumstances it is strategically best to join a confederation of the stronger parties allied to control the weaker just as the big fish in a pond will hang out together and gang up in order to eat the little fish. This is one of the principles of power politics.

If there are many weaker parties about the appropriate strategy for an aspiring leader is to bind them together into a single group united against the strong states just as when a flock of sparrows assembles together to attack a flying crow or owl. This tactic is one politicians easily understand because they are skilled at uniting special interest groups in order to win an election. If one examines the conduct of cartel members or of politicians trying to win office by election then it is easy to see some of these strategies in play.

Another strategy not mentioned is to refrain from fighting yourself but to maneuver competitor states to fight each other for your own advantage.

You can start to develop effective power strategies after evaluating the conditions at hand so that you can anticipate what will have a chance of success and what might result in failure. You must shape your strategies to take circumstances into account. As an illustration, we can look at the great legislative abilities of American President Lyndon Johnson who mastered the art of putting together the most unlikely coalitions in order to support highly controversial bills. President Johnson explained once how he secured passage of the 1964 Civil Rights Act:

> The challenge was to learn what it was that mattered to each of these men, understand which issues were critical to whom and why. Without that understanding, nothing is possible. Knowing the leaders and understanding their organizational need let me shape my legislative program to fit both their needs and mine.

One means to win strategic advantage is to mimic the martial arts of *taiqiquan* and use an opponent's strength against himself. The principle behind this strategy is that "strength" is equivalent to some characteristic in excess and that excess, through some angle, can also be viewed as a weakness.[48] Employing the converse of this principle the car rental company Avis once ran an advertising campaign which went, "Rent from Avis. The line at our counter is shorter (because Hertz Rent-a-car is number one)." A similar message: "Buy from us. As number two, we try harder."

Another effective strategy is to turn an enemy's weaknesses against himself. For example, in military strategy one tries to sow discord in an enemy camp when it's already disturbed by internal conflict. Another example is to support a leader's unqualified officials because they will continue to keep him in confusion. Napoleon is reputed to have said, "Never interfere with an enemy when he's in the process of destroying himself." You can profit from the folly of others especially if you don't disturb them when they are making mistakes on their own.

These strategies are built upon the Taoist principle that changes in phenomena run back and forth between the extremes of yin and yang. Since the extreme of yin gives birth to yang and the extreme of yang gives birth to yin, to bring about change in a situation you need only encourage a trend to grow to its extreme. Lao Tzu thus said, "In order to reduce something you must first stretch it. In order to weaken something you must first strengthen it. In order to destroy something you must first promote its flourishing. If you want to grasp something, you must first give it away. This means of acting is called profound insight." Hence Jiang Ziya was able to say,

> In order to attack the strong you must nurture them to make them even stronger, and increase them to make them even more extensive. What is too strong will certainly break; what is too extended must have deficiencies. Attack the strong through his strength.[49]

---

[48] Ronald Reagan, when questioned about whether his advancing age made him unfit for office, quipped during a debate with Walter Mondale that he refused to make a campaign issue about his opponent's youth and inexperience. This reply squashed the age issue and turned it to Reagan's advantage. When opponents depicted U.S. presidential candidate William Henry Harrison as a "cider-swilling rustic" due to his Indiana origins, Harrison turned this to his advantage by setting up log-cabin campaign posts throughout the country offering free cider to the voters and effectively painted himself as a "man of the people."

[49] *The Six Secret Teachings on the Way of Strategy*, trans. by Ralph Sawyer, (Boston: Shambhala, 1996), p. 46.

In developing strategies to obtain power and influence you must always carefully analyze the current conditions because this is the only way to determine how to act accordingly.

When there are many weak parties about the best strategy is to unite with the weak against the strong. Strength increases with unity so you'll be protected by the group and within it have the chance to gain power and influence. To demonstrate this, Mori Motonari, a Japanese daimyo of the Chugoku region, is famous for teaching the "lesson of the three arrows" to his sons, and this lesson of mutual advantage is still taught today in Japanese schools. Mori handed each son an arrow and asked each one to snap it, which they each easily did. Then Motonari produced three arrows and asked his sons to snap all three at once but none were able to do so. He explained that one arrow could be broken easily but three arrows held together could not, and having made this example he then urged them to work together for the benefit of the Mori clan.

If you want to be the supreme leader yourself in his situation you must be the first to stand up and harness the impulses of the group.

You must also work to decrease the size of your opponents as well as weaken the exceptionally powerful or influential who stand in your way. It is extremely difficult to become the supreme power if your peers are stronger than you so you must set about (secretly) to decrease everyone else's power as a means to increase your own strength.

Only the foolish dream of becoming the supreme leader within a group when there are many powerful competitors around. Master Guan warns that it is best to bide your time and accept that particular situation, to take no initiatives but to preserve yourself until the others show weakness or experience decline. In a fight between a goat and pig the ultimate victory only belongs to the butcher so you should preserve yourself and maintain your strength while others engage in a struggle.

In general, an individual can become the supreme leader if he takes the initiative when he only needs to triumph over small skirmishes or minor troubles. However, he will need to stand aside and preserve his strength if mighty dragons are contending for power. You can let them fight each other and after they weaken each other you can then come in and conquer them. In the James Bond movie, *From Russia with Love*, Blofeld talks with Russian defector Rosa Klebb and points to an aquarium with Siamese fighting fish, saying: "Siamese fighting fish, fascinating creatures. Brave but on the whole stupid. Yes, they're stupid. Except for the occasional one such as we have here who lets the other two fight … while he waits. Waits until the survivor is so exhausted that he cannot defend himself, and then like SPECTRE … he strikes!"

Taking the initiative is only wise when there are few opponents to contend with and they are all limited in strength. When many equally strong rival competitors are struggling for power it is best to let them weaken themselves in struggles while protecting oneself through non-involvement.

Whether in business, warfare or politics, you must learn to think strategically like this in order to know whether offensive or defensive measures are better for your affairs. Using the right strategy in the appropriate circumstances is a key for coming into power and maintaining it.

LESSON 25: Leaders Must Be Straightforward and Fair

> The ruler's mind should be straightforward but not severe. When conferring ranks and dignities on his subjects he should not exclude the worthy whom he selects by their talents and merits rather than by their seniority or status. This is done with the purpose of gaining even greater support so that he can become the only ruler of the world.

To become a powerful leader you must be fair and impartial in your decisions. You must practice fairness and justice where you rise above your own personal likes and dislikes. For the most part you must banish emotions from your decision-making.

Like General Dwight Eisenhower you should cultivate a harmonious demeanor and your speech should be correct. Your thoughts and behavior should avoid the extremes of severity because harshness will alienate you from your followers and peers. Righteousness should always appear at the center of your deliberations and while you should strive to follow correctness impartially you should not be offensive. In this way, everyone will be pleased because your manner will be acceptable to others.

You should also practice meritocracy rather than favoritism and promote the competent and worthy whether you like them or not. The famous executive John DeLorean wrote that he would promote men at General Motors he didn't like as long as they had talent and achievement. If you rely upon this policy then there will be loyalty, trust, harmony and accomplishment in your administration. However, if the undeserving are rewarded or honor is denied to the worthy then you will end up destroying your support and reputation, not to mention performance. When administrative and executive positions are not distributed justly the talented will eventually leave you or stop entering your service.

Good leaders therefore see to it that they fulfill the promises of reward and that the meritorious receive just recognition. A leader must promote those who are worthy without raising them to positions higher than appropriate and he mustn't ask the competent to do more than what they are good at. A leader promotes others because they are capable and deserving, which is the rule all that the people approve of in their hearts. Promoting others is to reward achievement while disciplining people is done to prevent further wrong-doing.

If you promote the right people then you will always be assured of tremendous benefit to your organization. If you promote the incompetent, however, then you will barely make up for troubles they cause and will have to spend untold effort to correct the mistakes they make. You must always be exceedingly careful about the promotions you make.

Rarely does a country or organization lack men of ability qualified to be promoted. Sometimes you must simply recognize the dormant power of those around you. It strikes me as strange, for instance, that the Bank of England had to go to Canada to find someone qualified enough to become the governor of the Bank (Mark Carney). Do you mean to tell me that the entire United Kingdom had no one qualified?

When the talented and competent are not used they will often leave so that they can apply their abilities elsewhere because they are like thoroughbred horses that want to race and be rewarded. Not only will this strengthen your competitors but over time this will reduce your own pool of top talent and necessitate that you work even harder to find the talents you require. You must reward those at hand, encourage them, motivate them, cultivate them.

The greatest leaders seem to be those who, upon finding people of talent, grab them, develop them and use them as did Winston Churchill, who made men be at their best. In selecting individuals for office a great leader shouldn't try to fill artificial quotas of gender or race etc. but should simply employ those who are the most worthy. Why should anyone bypass those whose wisdom, talents, virtues and experience are the foremost in a nation just because they lack the "perfect" sex, race, religion, or age? Political correctness is nonsense when the fate of the nation is at stake. The majority will approve when you use the best without restriction for this is the policy which lies in the best interests of the state.

This concern to employ the best should win the support of the people and ensure a prosperous fate. To become really great you must eschew political correctness and use the best without restriction as this is the policy offering the greatest benefit for the whole. Choosing "the best" may at times mean taking political correctness and diverse representation into account, but you must always remember that the real underlying principle is to employ the most capable.

LESSON 26: Leaders Must Employ a Simple and Honest Administration

The ancient kings, when struggling for control of the realm, emphasized the principles of straightforwardness and uprighteousness in government. They established regular and uniform measures throughout the realm and they governed the empire through simple rules and easy-to-understand policies. When governing and issuing decrees they would take into consideration the Way of Man [traditional standards and behavioral tendencies]. When awarding ranks and salaries they would reference the Way of Earth [local circumstances]. When initiating important business they would rely on the Way of Heaven [strategic timing].

The laws of a state must not become so complex that only the intelligent can understand them. Tax codes, for instance, should not become über-complicated. Effective administrations are straightforward in word and deed and employ simple policies absent of complications. The regulations should be very simple so that they can be easily understood by both the ignorant and intelligent alike and then they will be accepted by the public. When the rules and regulations are simple and concise the people will be able to obey them and they will be easy to enforce.

On the other hand, when a state imposes an overabundance of regulations this makes it difficult to follow any at all. If it's hard to follow the law because the rules are too numerous or conflict with one another then flagrant disobedience will be the rule everywhere, and then that idea will spread into other areas of society to the detriment of its overall welfare. Too many regulations simply ends up with everyone becoming a law breaker and a state then falls into discord and disarray. As the *Tao Te Ching* says:

> As the prohibitions and interdictions under Heaven become
> multiplied,
> The people will grow increasingly impoverished.
> As laws and edicts are increasingly publicized,
> The more thieves and robbers there will be.

While governments must create and administer the law for there to be justice in society a nation that focuses on the law rather than the underlying foundations of religion, ethics, virtue and morality will fail to produce a more perfect society from simply legalism. As Confucius noted, the legalistic approach of forbidding behavior focuses on the symptoms of a situation rather than the roots. When people focus on the letter of the law rather than its spirit there will be a diminution of moral responsibility which will become a cancer on the foundations of the state.

An overabundance of laws and regulations can make it virtually impossible for an individual to carry out even the most trivial of tasks without incurring extraordinary risk of some transgression. No one wants to live in fear that his tiniest action may lead to prosecution so excessive regulations never end up serving the citizenry. Problems must be approached by addressing the roots of a situation rather than its symptoms, so the best way to produce law and order is by promoting virtue and morality in the foundations of the national culture.

Taoist philosophy tells us that excessive legislation leads to disorder rather than order. Its existence openly advertises that a nation has failed in educating its people. Confucius taught that it was more important to teach the public a sense of shame than create an overly legalistic society. As he said,

> If you guide the people with laws and keep them in line through punishments they will stay out of trouble and have no sense of shame. However, if you guide them by virtue and keep them in line through the rites then in addition to having a sense of shame they will reform themselves.

The reason that too many rules and regulations is an administrative failing is because conditions reaching an extreme eventually produce the opposite extreme, which in this case would be lawlessness. An excessive amount of laws will produce lawlessness. Countries which impose an overabundance of rules on their citizens face the danger of transforming into despotic regimes because the rule of tyrants and dictators is characterized by excessive restrictions.

The laws of a land must therefore set the parameters of proper behavior but a nation must encourage public mindfulness of what is proper while avoiding a burdensome extreme on the citizenry. It is impossible to simply change the underlying fundamentals of problems by decree through the strategy of creating a bunch of laws. Blanketing unfavorable conditions with legislation is like putting cosmetics over acne—the situation may look good on the surface but it'll be festering underneath. Legislators must be avoid instituting laws that alter the external appearance of situations without actually solving the critical underlying problems because this can damage the situation even more at the same time you believe you are solving the problems and then move onto something else.

A wise leader therefore addresses the causes of situations rather than its outer appearance because placing a rock on top of weeds will only divert their angle of growth; the weeds won't die off unless you ultimately pull out their roots.

What people really want in life is to experience health, prosperity, peace and happiness; intimate and affectionate close family relationships; opportunities for an independent livelihood; freedom of personal expression where they feel authentic in their life; access to justice and protection when necessary; a sense of higher purpose for their life; and to get along in their life with little interference from the state. They want the ability to exercise their essence along a life of excellence with ample scope. When the government starts to needlessly interfere with people's lives by imposing itself here and there with unnecessary interventions then it has lost its social contract. What a disaster it would be if a government could gain control over your own money and supervise your purchases such as by imposing controls through a digital currency.

The best way to govern a country is to let the people lead the kind of life they wish without imposing unnecessary boundaries. The government sets a loose framework of laws and lets the people get on with their lives with very little interference from the state. It establishes fair laws and an impartial system of justice, and standardizes necessary definitions across the nation. Napoleon, for instance, established the Napoleonic Code of law and a system of uniform weights and measures in the territories he conquered so that uniformity was established across the land.

Adam Smith also believed that freedom of personal movement will bring about the greatest possible benefit for a country. As seen in the example of Hong Kong, where a state institutes a laissez-faire policy characterized by market freedoms and little government intervention, the people will prosper extraordinarily well under such conditions. You must recognize that a laissez-faire policy of no-policy is also a legitimate form of policy, that sometimes the best policy is to leave things alone, and the prosperity which seems to have no origins might be a government's greatest achievement.

Simple and straightforward policy, which addresses the necessary problems without doing the people harm, is the target of good rule. When laws and decrees are fair and convincing, when the public is left content in their place, when the ranks and salaries of officials are fair and well administered, and when important events are conducted in a timely and proper fashion, then a country will see order in its internal affairs and possess the circumstances necessary for prosperous advance.

Therefore, the laws and regulations of a country should reflect the realities of human affairs and current conditions. One reason that the Roman Empire stood for so long is because its administrative system was adapted to the local circumstances within its provinces. Thus it is that salaries and official posts must be coordinated with the local situation, and change to match with the times. Singapore solved its government corruption and talent problem in part by raising the salaries of officials to match those of business executives, thus making government service a magnet that pulled talent. Singapore tests its officials with larger and larger responsibilities to see if they have the qualities needed for the highest positions and in that way "gains insight" into its officials. It follows the advice within the *Three Strategies of Huang Shigong* to employ only the capable after first discerning who has ability.

When a country decides to undertake great affairs, Master Guan points out again that it should do so according to the proper timing, which is timing that takes into account the current conditions of the state. No one can ever underestimate the importance of timing because if your initiatives match the tide of affairs then it will be difficult for people to oppose you.

LESSON 27: Leaders Necessitate Disciplinary Measures

Whenever they launched punitive expeditions the ancient kings attacked only the states which were rebellious but not those states that were obedient. They did not institute attacks whenever some small incident occurred but only attacked those whose actions constituted a real danger, only those who over-extended their authority instead of those who merely failed to come up to standard. Within their own countries they ruled over the people with justice. Outside of their countries they summoned the other feudal lords using the influence of their political power and authority. The ancient kings would punish any neighboring states which refused to submit to them by seizing part of their territory. For those who were farther removed and disobedient the former kings relied on the use of military force to overpower them. When the feudal lords were rebellious the ancient kings responded by initiating military attacks and pardoned the lords when they yielded and became submissive. Both their military and civil responses were completely virtuous.

All leaders of organizations—such as an army, a business or nation— must ensure discipline among the members of their group. There must be methods to ensure that the people always follow your orders otherwise the occasional dissent or insubordination will terminate your leadership. Those methods are disciplinary measures, i.e. retaliation.

For this reason a political leader must respond immediately to renegade actions and sometimes use force to punish those who foster rebellion. A king will launch punitive expeditions against territories flaunting revolt, a military general will punish troops guilty of dissent, an executive will discharge individuals who refuse to follow instructions, and even the politician will find ways to institute retribution against those who oppose him. The possible means of retribution used are far too numerous to list. Leaders knows that the threat of punishment is an effective means for instilling discipline and order.

The real danger to your power is not some small breach of discipline by underlings but a serious conflict or standoff over a major issue. It is natural for any two parties, such as married couples, to disagree over issues without the risk of any long-term damage and there is no reason for undue concern in minor disputes because they naturally happen for everyone and are eventually resolved.

However, when a genuine challenge to one's authority arises Master Guan states that a ruler should definitely attack and punish opponents in such a way to "show who is the boss." A sovereign must always be careful when taking such initiatives because only the rebellious should be punished. Those who simply fail to come up to standard need only be reprimanded. For instance, governments often impose sanctions on foreign nations in order to punish the actions of their leaders but this only hurts their populations who never overthrow their leader, and is a poor strategy because it never produces change anyway.

Whenever leaders impose discipline on a group this bolsters their authority and reminds people of their power. However, while force can settle conflicts a forceful response will always draw the feedback of being questioned and criticized, and your actions may backfire. One must therefore be careful when exercising one's might because if a forceful response lacks the righteousness of justice then even a great leader can lose some of his support. Force must only be applied when righteousness is on your side.

The leaders of benevolent administrations have no reason to fear internal unrest when they are seen as truly working for the people. The contented rarely think of rebellion but start raising red flags when they can see that the government is not on their side, which seems to be happening in the world today as governments look like they are following some coordinated master plan as they commonly try to impose all sorts of illogical strange notions and unwanted restrictions on their populations.

Smart leaders spend time winning over their people, strive to pacify the alienated, and work to make their countries safer by disciplining antagonistic neighbors when they get out of line. You do this by withholding or taking away what is dear to those challengers such as by imposing tariffs or quotas that hit them in the pocketbook.

The actual use of force is usually reserved for those stationed much farther away because a leader tries to maintain stability on the home front while keeping aggression at arm's length. The highest method is to use force to threaten distant states rather than engage in actual fighting, and thus accomplish your aims while preserving your strength. Neighboring territories can be a source of either alliance or discord, and it is better to keep them as friends rather than as rivals.

When fundamental conflicts do exist between parties a leader's first strategy should be to use the soft to overcome the strong. In politics, this means that he should attempt to peacefully influence other lands through cultural and economic means rather than through the bullying of aggressive force.

If a political leader must actually employ military force he should make sure that there are legitimate reasons for this response and must effectively communicate these reasons to his people, his peers and the offender. When a leader is honest and trusted by his foreign allies because of this then this respectful deference will go a long way toward protecting his standing among them despite his use of forceful ways. President George Bush followed this principle by broadcasting his position to the public and contacting all his allies to form a multi-national coalition before he militarily attacked Iraq for intervening in Kuwait.

LESSON 28: Allies are Needed

There are certain conditions which can make one important or unimportant and which produce strength or weakness. When the feudal lords are cooperatively united they are strong whereas they are weak when they are separated. With all its capability even a fine thoroughbred is bound to become exhausted when challenged by a hundred ordinary horses in succession so of course even the most powerful state of its time is bound to be weakened when attacked by states from all corners of the world.

It is no secret that those aspiring to politically lead the world will face opposition when people sense a threat to their own interests, and it is inevitable they will from time to time see various groups united against them. Therefore you need to form or join alliances and coalitions; you need to cultivate dependable friends and trustworthy alliances for your support. To lead you must be involved with others, and involvement translates into alliances.

Strength derives from cooperation with others while weakness comes from isolation. Alliances will help protect you and allow you to conserve your strength so that you can direct it toward more important objectives. Two or more united nations will be stronger than a single country so allies and alliances are important to gaining power.

Just as soldiers will tire of repeated battles, even the strongest leader in the land will become fatigued if he must singlehandedly fight opponents without end. America, for instance, would find it impossible to shoulder world security concerns by itself if it had to fight battles on multiple fronts simultaneously. This is why it is always trying to split the unity of China and Russia because when united these world superpowers could create troubles for America on multiple fronts.

To assume influence over the world's states a supreme leader should gain the support of various groups either by rallying them behind his own banner or by joining their pre-existing alliances and working to gain influence from within. Britain uses a strategy of encouraging nations to join its alliances and then slowing gaining control over them through that membership.

National isolationism, as can be seen in the histories of China, Tibet and Japan, typically results in countries falling far behind their peers and endangers the security of a nation because others are typically always advancing themselves with new economic and military technology. You have to keep up with modernization in all forms and isolationism prevents this.

As a leader you must learn to deal with others you don't even like. Professional negotiators say that if you refuse to deal with enemies, competitors or contenders by avoiding any form of dialogue then you'll never be able to influence them at all. Hence there is an importance to keeping open the lines of communication and remaining on good terms with every kind of state.

When the appropriate conditions change, such as warfare or the discovery of a rare natural resource, a state that seems unimportant today can assume strategic importance tomorrow. Therefore, a leader should try to keep on good terms with as many parties as possible.

LESSON 29: Strategies for Survival and Influence

A strong state gains by winning the loyalty of small states but fails if it becomes too arrogant. A small state gains by temporarily yielding to the big powers but fails if it breaks with its strong neighbors. Countries, whether large or small and strong or weak, each have their own particular circumstances and individual strategies for ensuring their preservation. Subjugating the neighboring states through influence and using force to threaten distant states are the methods employed by the country of a king. To unite the smaller states in order to attack the larger is the method used by rival states. To maneuver the small states on one's borders to attack other small states on one's borders is the means used by a central state. To conserve its resources and humbly serve the strong so as to avoid offending them is the means used by a weak state. Since time immemorial no state that took initiatives against its circumstances at an improper time has ever been able to keep its name and standing. There has never been a state that did not suffer defeat after having encountered difficulties and gone against the times in attempting too large an alteration of its conditions.

Evolutionary biology demonstrates that there exists an appropriate survival strategy for every type of animal that is prey for other beings, otherwise it wouldn't exist. The same holds for every type of nation whether it be landlocked, large or small in size, bereft of resources, sea faring and so on. Furthermore, since these survival and growth strategies are specific to the nation it would be quite inappropriate to expect all nations to be following the same developmental model or even use the same form of government. Homogeneity would eventually kill off a forest and a one world government would end up destroying prosperity and freedom in the world.

Nature shows that "uniformity of the gene pool" does not work in the long run. States should structure themselves in light of their fundamental conditions and character for were a junior to act as a senior and senior as a junior then a nation might cause its own destruction. Mongolia, for instance, is land-locked by both China and Russia. It depends upon Russia for nearly all of its energy needs and upon China for its export markets. If it tried to assert itself as the pre-eminent one of these two it would certainly be slapped down and cause itself some ruin. It has no choice but to allow Russia to run its natural gas pipelines across its territory into China.

World history reveals many strategies which nations have evolved in order to ensure their survival. Island nations, for instance, tend to develop strong navies and become excellent traders. England falls into this category. Landlocked nations with few geographical advantages, on the other hand, have tended to develop strong, maneuverable armies as well as expertise in the arts of diplomacy. Germany is not landlocked but falls into this category. Geopolitical strategists sometimes say that Germany's aggressive nature is because it is mostly surrounded by other countries, and this has caused it to develop the reflex of acting first militarily in order to preserve itself.

Smaller states have tended to preserve themselves by showing deference to larger nations while larger nations, to protect their borders, have learned to show magnanimity toward their smaller neighbors and use them as buffer zones against invasion and attack.

As Lao Tzu said,

> When a large nation shows deference to a small nation it wins over the small nation. When a small nation shows deference to a large nation it easily wins its trust. One uses deference to make the small nation seek shelter in the large nation. The other uses deference to make the large nation magnanimous towards the small. What the large state desires is to lead the small states. What the small state desires is to join and work with the large state. Since both get what they want, it is fitting that the large state be humble.

In offering its smaller neighbors shelter and protection a larger state safely increases both its prominence and effective territory without undue cost nor effort. This is not just a matter of benevolence but a matter of strategic interest. From a purely geopolitical point of view friendly relations with smaller neighbors can create buffer regions to invasion, just as North Korea does so for China, while increasing a nation's power at little risk since smaller neighbors treated kindly usually act in accordance with their larger host's wishes. Friendship offers great strategic advantages.

Every type of state has refined a particular strategy for self-preservation just as evolution has led to different animal species developing their own specific survival mechanisms. These strategies for preservation (security) and prosperity not only include a state's political system, economy and internal organization but also its external relations as well.

A nation is no different than a living organism in that it must adapt itself to new circumstances in order to avoid destruction. To ensure survival even individuals must follow this policy[50] because to go against conditions is dangerous. In short, the strategies you use for ensuring your nation's survival should be based upon your circumstances.

All states must also develop stratagems for increasing their external influence because relations with other nations affect one's prosperity and risks of survival. The premier state augments its influence by winning the courteous respect of close neighbors while making more distant states fearful of its military capabilities. A lesser state unites with other smaller powers through alliances to increase its effective power and ability to

---

[50] For instance, the Biblical King David at one time feigned madness and the Roman Emperor Claudius feigned uselessness to remain out of harm's reach. The famous Buddhist Abhidharma master Vasubandhu also feigned madness in order to gain access to spiritual teachings.

preserve itself against larger states. Some countries, such as Israel, protect themselves by maneuvering their neighbors into fighting amongst themselves thus hoping to weaken them endlessly.[51] A weak country, unable to do anything, will be found making concessions everywhere just to stay alive.

If nations do not keep up with the modernization trends of the times they will fall behind and open themselves up to invasion. Ethiopia, for instance, was able to hold off an invasion by Italy during the First Italo-Ethiopian War only because it had stayed current and had purchased modern artillery. Modernization is fearful and sometimes painful to parties who must change so the key is not to alter their conditions too fast but give them time to adjust.

It is said that the Shah of Iran, for instance, lost his country in part because his modernization policies forced Iranians to radically alter their conditions too quickly and this then gave rise to internal strife. He failed to convince his people of the benefits of change and give them sufficient time to digest those changers. The Kingdom of Bhutan, on the other hand, has changed its form of government several times and before instituting its latest variation (as a constitutional monarchy with a parliament) it went out of its way to convince the people of the benefits of this change.

We must embrace change, but should not have to swallow too much at once because it takes time to get used to a new situation. Only after adjustment will it be easier to continue everything in the same way as before. The Shah of Iran, despite the great things he accomplished, violated this principle in too frequently trying to make large changes in one leap. The public soon becomes exhausted from having to make one new change after another without being able to forge any type of harmonious relationship with the rapidly changing situation.

The rule to follow is that a leader will not be able to retain his authority if he makes too significant demands on his following. Abraham Lincoln, although opposed to slavery, did not immediately grant the emancipation of slaves upon becoming president because he concluded that it was too radical a change and would have further divided the nation (by driving border states to secession) so he left this proclamation to a more appropriate time.

When a nation's people cannot digest the pace of change imposed

---

[51] This is a strategy that wise statesmen of every country need to be wary of—that another state manipulates your own country to your detriment for its benefit. The vital principle is to avoid acting rashly in international affairs, and then you will have a better chance to avoid becoming entangled in matters adverse to your well-being. "Statesmen" earn their title from calmly viewing world affairs to see how their state will benefit or lose, and then acting accordingly. They never rush into matters, unless they are well-prepared ahead of time.

upon them, perhaps a pace demanded by the requirements of modernization, this can endanger a country's foundations so a leader must be careful of what he demands from them. Some countries, like Oman, were able to do this quite successfully.

Countries which try to radically alter their conditions too quickly may cause internal unrest and upheaval because radical changes often gives way to reaction.[52] Accordingly, the wisest states will avoid excessive change when trying to improve their circumstances because this equates with excessive risk. They will plan their development strategies to involve digestible objectives and will cautiously implement their plans using appropriate timing.

LESSON 30: Overthrowing a Leader

> Those individuals who wish to overthrow their sovereigns and conquer the realm [for themselves] should not depend solely on military force for victory. Instead they should first be certain to make proper plans, occupy advantageous terrain, proclaim a doctrine everyone can understand, develop close relations with allied states, and then wait for the right moment to act. These are the methods that will make one a king.

Plans to conquer a country purely through military means are always insufficient because you must win over the people – both the leadership elite and common man. This is why intelligence agencies try to launch "color revolutions" within countries in order to overthrow their governments and install their own. While a country may succumb to the military might of another, force is a poor glue for holding together a realm because even a victorious conqueror must win the support of the people in order to consider them "his people," otherwise his reign will always be on insecure footing.

Alexander the Great, when trying to conquer Persia, tried to politically inject himself into the Achaemenid royal contest as a better candidate than Darius III and thus he was able to win over many conquered Persians. He set out to prove that he was more worthy than Darius III, seized the majority of the strategically important cities of the empire, and rallied the Persian nobility to his cause by not relieving them of their rule in order to

---

[52] A reformer must remember that his greatest opposition will come from those who profit most by the present order. Therefore, leaders must be careful of moving too far too fast. Too many leaders have destroyed themselves due to lack of patience and by intervening in affairs which were well enough left alone.

prevent any future rebellion, insurgency or opposition. At the core of his approach to politically running his empire was the maintenance of existing local arrangements with slight amendments.

Therefore a political leader works to win over the people's hearts before he takes action. He propagates a doctrine that everyone can understand, which resonates with the people's will, which promises them advantage, and which speaks to their hearts. When a leader speaks to the people in their own language and articulates their true concerns he will win support from all areas of the country.

Whatever their vision a revolutionary's message must be righteous because there must always be strong justification behind establishing a new order. The need for the legitimacy of change is so deeply imbedded in our psyche that we still pay reverence to centuries-old documents such as the Magna Carta, the Declaration of Independence, or the Oath at Muye where China's King Wu revolted against the Shang dynasty. In the lesser known[53] story of the Oath at Muye, which has been left to us in China's *Book of Records*, China's King Wu called for an overthrow of the despotic emperor "Zhou the Terrible." His story of political misrule can be found throughout the hexagrams in *The I Ching Revealed*.

King Wu claimed that Heaven had mandated him to punish Zhou for his tyrannical wickedness, cruelty to the people, and his betrayal of the traditions of the earlier Shang rulers. Technically this was an act of rebellion but centuries later Mencius remarked,

> Men who abuse humanity are crooked; men who abuse righteousness are villains. They are both merely vulgar men. I have only heard that King Wu killed a vulgar man named Zhou, but I have not heard that King Wu ever killed a sovereign.

Since Mencius talked this way countless generations after the event it only supports Master Guan's observation that the memory of righteousness runs long and deep in a nation's psyche, and that the common man will seek righteousness in their leader's actions and the actions of other nations. When righteousness is abandoned the public will withdraw its support from a leader and his endeavors. On the other hand, you can make use of public opinion in accomplishing your aims if your goal contains the message of righteousness and advantage and incorporates a sense of justice.

Though you may be a very popular leader who easily wins public approval for your actions you must never forget to make adequate plans

---

[53] That is, lesser known to non-Asian readers who are more familiar with the stories behind the French and American Revolutions as well as the history behind the Magna Carta or Swiss Confederation.

and preparations for your goals. You must also never forget the importance of properly sizing up a situation and opportunely timing your actions because nothing is more effective than striking an opponent, or setting out on a new initiative, when the timing is most appropriate. If your plans and support are in place and you act with opportune timing then you can successfully become the supreme leader of a realm.

As *T'ai Kung's Six Secret Teachings* states this factor of timing is crucial. There must not only be adequate reasoning but auspicious timing as well before you even dream of succeeding in such an important task as overthrowing a sovereign:

> If there are no ill omens in the Tao of Heaven, you cannot initiate the movement to revolt. If there are no misfortunes in the Tao of Man, your planning cannot precede them. You must first see Heavenly signs, and moreover witness human misfortune, only thereafter can you make plans.[54]

To the orthodox the unorthodox is only palatable during times they want to escape danger. If you try to institute radical changes without such timing you may doom your chances of change. As a result, today U.S. military intelligence uses supercomputers to game out possible changes in American society due to various policies and advises actions only when certain population statistics reach particular levels.

Normally books like *Culture, Country, City, Company, Person, Purpose, Passion, World* that contain all sorts of economic and socioeconomic cycles are used to help leaders know what to cyclically expect in the future. *Pendulum*, by Roy Williams and Michael Drew, and the works of William Strauss and Neil Howe give insight into generational forecasting. Books like *Cosmos and Psyche* by historian Richard Tarnas and *Planetary Cycles: Mundane Astrology* by Andre Barbault provide astrological timing that leaders can use in forward planning, an example being the possibility of a new era of Japanese prosperity due to a Uranus-Pluto trine at the end of the 2020-30 decade but then challenges during 2036-37.

In *The Value of Astrology* Barbault notes that Saturn-Neptune aspects often produce public revolts or revolutions and frequently radicalize in the style of the extreme left whereas Sun-Jupiter conjunctions often mark time periods when you can achieve an armistice (peace agreement) as a consequence of a state of mind that can last for a certain time. *You can use yearly Sun-Jupiter conjunctions to try and forge peace agreements of all kinds!* Wise leaders who know of such timing can take advantage of such tendencies to

[54] *The Six Secret Teachings on the Way of Strategy*, trans. by Ralph Sawyer, (Boston: Shambhala, 1996), p. 52.

push for peace, which is Master Guan's advice on auspicious timing. This astrological tendency for peace agreements or détente is like a verb without a predicate that can be attached to many situations. This annual factor can be applied in many ways by the smart leader according to what is happening in the world at that time.

Economic revolutions, on the other hand, are linked to a trio that includes Jupiter and Saturn cycles with that of Uranus with the Uranus-Neptune cycle as the framework. Basically, the right moment to act for *big things* therefore involves contemporary conditions as well as the patterns of the heavens.

The wisdom of the Zen master is that he will patiently wait for the perfect time to kick someone in order to propel him to some level of realization. Sometimes a doctor just cannot reform a patient's unhealthy habits until he is deathly sick and finally amenable to changing his lifestyle. Hence at certain times refraining from action and waiting can be the highest form of skillful means.

LESSON 31: Punitive Expeditions

When the ancient kings engaged in punitive expeditions they made sure that there was justifiable cause for their attacks, which they only directed against cruel and violent states. They examined the conditions to determine what was possible, and made up their minds on whether to launch an attack by carefully estimating their strength and by studying the circumstances to decide on the proper moment for taking action. For this reason, when the ancient kings conducted their punitive expeditions they first engaged the enemy in skirmishes before launching a full-scale attack and they first undertook full-scale invasions before seizing territory.

A successful attack upon an enemy state depends upon having both justifiable cause and proper preparations. When one has justifiable cause he'll also have the full support of both the public and military forces. That's why the *I Ching* says, "On expedition with good cause the general will have good fortune."

As stated earlier, when the leader of a state acts with good cause then the people will follow his orders and his military forces will have high morale. But if you do not make careful plans or preparations before dispatching troops into foreign territory, such as understanding the geographical features ahead, the odds for defeat rise tremendously. Without careful preparations an entire army can be carelessly destroyed.

The importance of deeply planning strategies and tactics will help to bring success, and the task of scripting one's strategies has a memorable emphasis in *The Dirty Dozen* where the soldiers in the movie repeat a rhyme to help them remember their roles in a coming attack: "One, down to the road block we've just begun. Two, the guards are through. Three, the Major's men are on a spree. Four, Major and Wladislaw go through the door. Five, Pinkley stays out in the drive. Six, the Major gives the rope a fix. ..."

Only astute observation and accurate judgment can enable military commanders to decide what forces and tactics to employ. In warfare it is therefore often beneficial to first test one's adversary by engaging in small-scale assaults and skirmishes before initiating major attacks. This is a prudent way to judge the enemy's forces and reduce the costs of battle. In the commercial field of marketing, companies test market products before committing to them as well and thereby reduce their risks of putting substantial funding behind products no one wants.

No matter what the size or scope of military engagements you must not fail to take advantage of favorable timing and circumstances to win. To initiate a war on your own when the opportunity is not right or when you are not well prepared is simply suicidal. When you commit troops without prior planning and preparation you are recklessly endangering lives and risking your nation. Such behavior is disgraceful simply because it is avoidable.

In trying to win the world you should always calculate the advantages and disadvantages, opportunities and difficulties before initiating any major undertaking. You should also always analyze your opponent's moves with utmost concentration. In particular, you should initiate aggressive moves with prudence and only after careful consideration.

LESSON 32: Leaders Must Calculate Their Own Strength

> Those mastering the art of attack would calculate the number of troops needed to match the enemy's force, the amount of food needed to match the enemy's supplies, and the quantity of equipment needed to match their opponent's equipment. When comparing troops, if the advantage in numbers lay with their opponent they would not attack. When comparing food and supplies, if the advantage lay with their opponent they would not attack. When comparing equipment, if the advantage lay with their opponent they would not attack. Those skillful in warfare avoid the enemy's invincibilities to attack his vulnerabilities, his strong spots to attack his weak spots, and his

difficult targets to attack his easy targets.

To determine whether or not you can win a conflict through force you must employ calculations, which was a famed trait of the great military strategist Zhuge Liang during China's Three Kingdoms period. You must accurately compare your own strength and abilities with those of your opponents and you must delve into the details where the details matter most.

In the military field you must compare equipment and resources, soldiers and their maneuverability, commanders and their tactics. You must be able to judge whether your supplies, your men or your equipment will be able to last long or not because you should launch a war if your resources will prove insufficient. If your measurement of powers is not thorough then you will not be able to determine who is stronger and who is weaker. Once you find a weakness on your own side you have to shore up the deficiency.

Entering battle is always a matter of taking calculated risks so you should enter into (self-initiated) battles only when the risks are in your favor and the potential rewards are greater than the potential losses. You can accurately discern those odds only if you can master the art of comparison.

Whether you study the works of Clausewitz or Sun Tzu for business,[55] military or political concerns you will always hear that you should attack an enemy at its weakest points and you should be stronger than your opponent at the decisive times and places. However, to attack your enemies at their weakest points requires that you first master the art of comparison and calculation to know where these places may be.

LESSON 33: Good Management Relies upon Moderation

To have a state firmly in control one does not need to strictly revere ancient tradition. To order the whole realm one need not be a master of military might. Becoming a lord protector or king does not lie in petty details [such as perfecting codes and regulations]. Rather, if one does not act properly then the state will be endangered. If one uses excessive force then his political power will collapse. If one is inclined to act rashly then the

---

[55] Al Ries and Jack Trout have published a number of short but excellent books on business strategy which embody such sagely advice. *Marketing Warfare* is especially recommended as it shows how to apply the principles of military warfare to marketing strategy.

result will be disaster. On the other hand, careful planning will bring great results. If one's endeavors are successful then fame will follow. If one has delegated power successfully his orders will be carried out. Such is the grand scheme for ensuring security.

Success in governing states, or in being able to restore order to the world, does not depend on blindly following ancient ways. Adhering to ancient traditions constitutes a fatal flaw of sub-optimal solutions, and your predictability can be used against you by any alert opponent.[56] To blindly follow traditional ways is a poor means of governing because you become excessively rigidified when reforms are necessary. For instance, many Muslim nations that follow the *sharia* legal code refuse to acknowledge that society has evolved and it has never produced the prosperity they consider promised to them. In today's world you need to master science and technology to succeed instead of simply following old religious texts and believing that a theocratic state, like Communism, will solve the people's problems.

When you come into power you must certainly abide by certain traditional norms, but you must also reform your state to match with the needs of the times. Laws and customs develop in response to the needs of the times. Whatever is inappropriate from former days should be abandoned while whatever is good from former regimes should be adopted, and whatever is good for today's world should be employed as well. The basics that have proved sound in the past and stood the test of time should not be changed unless absolutely necessary. This is common sense. Communist Russia at one time tried to outlaw marriages and of course this disastrous policy soon had to be reversed.

The *Three Strategies of Huang Shigong* tells us that you need to act according to circumstances and avoid responses based on imagination, memory of the past, habits acquired in other circumstances, or tradition. Just rely on observation of the present circumstance and create your plans in response.

Chinese Taoists have noted that there has never been any constant in traditions or culture so a leader should formulate manners and culture without being ruled by traditions and culture. However, whenever possible *development* reforms should be in tune with the core positive DNA of the

---

[56] One of the reasons a Chinese emperor is called a dragon is because he never reveals his true plans. Since you can never guess what he's thinking he mirrors the sublime inscrutability of dragons. He's subtle to the point of formlessness and mysterious to the point of soundlessness. An emperor becomes powerful because of his unpredictability and uses this strategy so that no one can second-guess him.

culture.

Therefore a nation's policies must be flexible enough to head left or right as appropriate to match with the times. Singapore is an excellent example of a state willing to reverse its prior policies because they were found errant or new ones were considered an improvement. An inchworm will sometimes move backwards in order to push forward so one shouldn't consider it a failure to reverse policies at times. The same goes for your positions on issues because if facts come to light that negate your old position then you should change it, and if newer facts come to light that negate that position then you should change your position again. There should be no embarrassment in changing your position, but embarrassing in refusing to change when new facts come to light.

To govern wisely, whatever is no longer appropriate from former regimes is to be abandoned while whatever is deemed beneficial should be adopted. This is the principle behind positive innovation and a means by which businesses develop themselves to become much more profitable. You should try to benchmark all your processes and adopt what is best no matter the origins.

One measure of a powerful political leader is whether he can actually institute the changes necessary for the welfare of his state when the required changes go against historical tradition. Although the job of instituting reform in the face of heavy opposition may seem like a heavy burden, leaders must remind themselves that each and every one of the "ancient ways" was at one time an innovation that encountered resistance in its day as well.[57] Just as those innovations have now become part of fixed tradition so in time will the present necessary initiatives become accepted.

One example of a much resisted innovation, which earned the merit of a thousand years, was the time Zen master Baizhong Huaihai broke with tradition and created a new form of Zen monastic life in China. Rather than allow Buddhist monks and nuns to maintain themselves based on donations he insisted that they establish self-sufficient communities based on their own agriculture and self-work. Though criticized for his efforts and reviled by some it was only because of these changes that the Zen school was able to survive the Great Anti-Buddhist Persecution instituted by Emperor Wuzong when other forms of Buddhism in China were attacked. Going against tradition, even though it was the correct adaptive response to the contemporary situation, earned Master Baizhong only scorn, criticism and resistance even though his behavior was an exemplification of the highest foresight because it saved the Zen school

---

[57] In *Analects* XIII.1 we find, "Tzu-lu asked [Confucius] about government. The Master replied, 'Lead them; encourage them!' Tzu-Lu asked for further advice and the Master responded, 'Untiringly.'"

from extinction.

A true leader always maintains a vision of the larger picture, which is to ensure the survival and prosperity of the state. Hence those who rule a country must always think deeply before they create new policies and directives. When it comes to guiding a realm the major objectives are important rather than the trivial details. Since the decisions a leader makes will ultimately affect the fate of his nation, short-sightedness in focusing on the nonessential, or in blindly following ancient ways, might endanger the state.

Acting in accord with present circumstances, even if it means going against tradition, is wisdom even though getting people to accept new ideas for their own benefit is often an uphill battle. Since most people resist change the best strategy for instituting transformation is to skillfully and invisibly guide current trends towards what you want to happen rather than trying to forcibly institute revolutionary changes in one fell swoop. Hexagram 49 of the *I Ching* (see *The I Ching Revealed*) explains how to first drum up support before trying to institute revolutionary changes in a populace. You must be careful not to force on people too much at once, especially if it strikes the cord of unfamiliarity.

Unfortunately, the short governing tenure of most world leaders and their prior lack of experience in long-term geopolitical designs combine to make this a most difficult task. This is one of the reasons why it is first necessary to promulgate a simple doctrine before instituting radical policy changes, which is because a successful leader cannot rely on using too much political or military force to achieve his objectives.

When a government forcefully imposes too much upon the people, or institutes inflexible policies that cannot change with the times, a state will in due course weaken. When policy doesn't match current conditions the foundations of a state will weaken because of consequential inefficiencies and friction; when policy is too hastily contrived, showing a lack of careful planning, it will also evoke excessive public reaction.

Basically, as stated in *The Master of Demon Valley* (an ancient Chinese book on statecraft),

> In the affairs of rulers, the general public, or important people, if something lofty and noble is possible and appropriate, decision is made in favor of that. If something that does not require expenditure of energy and is easy to accomplish is possible and appropriate, decision is made in favor of that. If something that requires effort and intense exertion but cannot be avoided is possible and appropriate, decision is made in favor of that. Decision is made in favor of that which eliminates trouble, if possible and appropriate; decision is made in favor of

that which leads to good fortune, if possible and appropriate.[58]

In conclusion, if something is beneficial to the people it need not follow ancient ways. Administering policy is like dispersing medicine—it is done with care, with an appropriate explanation, and according to the needs of the patient.

The supreme leader thus carefully formulates a grand strategy for his country and then campaigns for, rather than forces, the acceptance of his policies. When leaders promote balanced policies designed with the people's benefit in mind then the people will gladly follow their directions without need to be reminded of the government's authority.

In short, force is a terrible way to make people become followers because the best way to attract bees is by using honey, namely incentives. When the goals are lofty governments should use rewards as an inducement to change rather than just punishments. Thus you should use public relations campaigns to move society in the direction you want before instituting those changes. When you can establish gradual change through doctrine and have the people submit gladly, rather than threatening them with punishment or ostracism, then the state's foundations will not be endangered.

The United States is trying to force feed an unwanted agenda of LGBT issues, climate change, unsupervised foreign immigration, digital currency, ineffective vaccination, anti-patriotism, anti-Christianity, weak families, Russia phobia, controlled living spaces, and the abandonment of fossil fuels on an unwilling, wiser public so it is only a matter of time before we will see whether this onslaught damages the foundations of the state.

LESSON 34: Strategy, Circumstances, Timing and Power

States that are competing for dominance must first strive for good strategy, favorable circumstances, and power. The ruler's happiness or unhappiness depends upon the quality of his own strategies. The situation or circumstances are what cause his country to be considered unimportant one moment and taken seriously another. The willingness of his army to take orders to advance or retreat depends upon the ruler's authority. Therefore, if he masters strategy his ambitions will be realized and his orders carried out. If he masters conditions and

---

[58] *Thunder in the Sky: On the Acquisition and Exercise of Power*, by Thomas Cleary, (Boston: Shambhala Publications, 1993), p. 51.

circumstances he can seize the territory of large states and surround the military forces of strong states. If he masters power he can overcome the military forces of the realm and summon the feudal princes to court.

The leader who becomes known for accomplishments is an individual who never trusts his fate to luck. He knows there are certain key elements to making an organization strong and so he develops those characteristics in every possible way. Such rulers formulate good strategy, promulgate law and order within their country, cultivate economic, political and military power and then wait for favorable circumstances before acting.

Napoleon serves as an example for this lesson because he was always careful in selecting his tactics, promulgating a code of law to regulate the people, emphasizing his authority, planning his moves and waiting for the right time to act.

The leader whose plans becomes foremost bases his actions on sound preparations. First he derives good strategies, next he waits for the appropriate circumstances to act, and then at the right time he uses his power to act. Strategy, timing and then proper execution—these three enable leaders to win their objectives.

When individuals master strategy, the evaluation of circumstances and the use of propitious timing—so that they can use the right strategy in the right circumstances at the right time—they can become an unstoppable force. These capabilities will empower a world protector so that he gains influence over every state and nation.

LESSON 35: Leaders Must Recognize the Designs of the Time

The divine sage king examines the condition of the various states of the realm in order to know the strategies and designs of his time, to determine the deployment of military forces, to apprehend who could win what territories, and to comprehend the result of various decrees.

To be able to influence the world you must be fully aware of the geopolitical desires of every nation and the possible strategies they will employ to achieve their individual objectives. The heads of state receive reports on this from their intelligence agencies and state departments,[59] but

---

[59] These reports are usually biased and tend to have an agenda. Even so, France is particularly good at providing their leaders a good overview of geo-conditions and the background of leaders their politicians will meet while the United States is

I found that the Youtube channel WhatifAltHist can help train people to start thinking this way.

You must understand the unique political, economic, historical, geographical, social and military conditions represented by every country, and should be able to predict how each country will react under various circumstances.[60] If you don't understand a nation's psychological, social, historical and economic conditions then you won't be able to understand its various changes and developments.

You must be capable of comprehending all these factors and of thinking independently. In short, you must be able to see what others cannot see and understand what others cannot understand. Only then will you be able to act decisively when the timing is appropriate. As Zhuge Liang once wrote, "When opportunities occur through events, but you are unable to respond, you are not smart. When opportunities become active through a trend, and yet you cannot formulate plans, you are not wise. When opportunities emerge through conditions, but you cannot act on them, you are not bold."[61]

By understanding the world's conditions you will know how to use your diplomatic, economic and military forces and how to increase your country's power. Even if you have knowledge of every state, however, you will not be able to conquer the world if you lack opportune timing and good tactics, which are dependent upon both the external circumstances and having excellent advisers.

Whether it be geopolitics, business or military affairs, the participant who has no knowledge of the overall situation, especially his opponent's conditions, will be at a loss in deciding what to do. To become a supreme political leader you must have absolute confidence about the decisions you and your opponents will make. For instance, to initiate a military campaign when you lack intelligence of an enemy's situation would be foolish. Whether your objective is victory in war or success in another field a knowledge of the playing field is mandatory. Just as a statesman strives to understand territorial alliances and a general strives to comprehend the directions in which opponent forces are likely to attack, assessment is crucial to formulating sound strategy while strategy is the cornerstone of success.

---

abysmal.

[60] This is a most proper task for a government's department of state (ministry of foreign affairs) and its intelligence agencies. A nation's foreign relations, whether during war or peacetime, should always rest upon accurate strategic intelligence.

[61] *The Art of Wealth*, trans. by Thomas Cleary, (Deerfield Beach, Florida: Health Communications, 1998), p. 260.

In the field of geopolitics, no matter how good your knowledge is of every state and country you could not possibly conquer the world without good strategies, tactics and timing. The supreme leaders who aspire to gain power and "conquer" the world through force or influence therefore strive to become aware of the whole situation. To this end, over the years they practice developing their insight into psychology and trends just as Mao Zedong read *Dream of the Red Chamber* five times in order to gain insight on how to deal with certain types of people.

I would personally recommend *The Romance of the Three Kingdoms* for gaining knowledge of the strategies countries will employ against your own and watch the Chinese movie version (with English subtitles) available on Youtube as "Three Kingdoms (2010)." A particularly famous quote from the *Three Kingdoms* is its first line: "The empire, long divided must unite, long united must divide. This is how it has always been." *The Thirty-Six Stratagems*, another Chinese classic, is also highly recommended.

You should particularly study how Britain establish colonies in the past, controlled their populations, and how it attempts to manipulate other countries today for its geo-political purposes, some of which are masterminded in the City of London. The history of Britain's East India Company and how it gained tremendous influence in India with so few personnel is worth studying.

Supreme leaders of the past who aspired to gain power gain knowledge of the situations at hand, used the appropriate strategies, gathered allies, commanded their military forces properly, understood how to distribute the territories they conquered or benefits they gained, and promulgated the right decrees. In what facet were they found negligent?

LESSON 36: Leaders Must Never Employ Force Improperly

If a ruler uses military force to attack those whom he detests simply in order to benefit himself he will elicit the disaffection of his neighboring states. However, if his power is directed against those whom everyone abhors and little profit accrues to himself then he will become strong.

A ruler must never use his position of power for personal gain or egotistical initiatives such as for avenging personal slights. When he uses power against an adversary it must be for unselfish reasons such as to save the people by punishing an aberrant and bringing peace to the world. The wrong use of power is the common pattern we see today where a sovereign nation is invaded, the invaders pilfer is assets and destroy its infrastructure, they then fund its reconstruction but the elites steal the majority of those fund sin various ways.

History is replete with cases of punitive expeditions. Gross injustice demands a strong response. The 2022 Russian invasion of Ukraine had multiple reasons but is primarily described by some as a punitive expedition to eliminate bio-warfare labs and save the Russian speakers of the Donbas who were being exterminated in a genocide. In 518, Negus Kaleb of Axum (modern-day Eritrea and Ethiopia) launched a punitive expedition against the Himyarite Kingdom in response to the persecution of Christians by Himyarite Jews. The 1979 invasion of Vietnam by China was in response to Vietnam's invasion of Cambodia, and Deng Xiaoping commented that "Children who don't listen have to be spanked."

If a leader acts with selfish ulterior motives then even if he were to destroy a state that everyone detested he would generate resentment and lose the world's support. The many military interventions of the United States to change regimes, and how it has militarized its own currency has put it into this very position.

Therefore, a leader never assaults others just to show his power or to deflect the attention of his nation away from internal problems. Righteous behavior in the use of force always attracts adherents and it strengthens a leader's authority and standing.

LESSON 37: The True Measure of a Leader's Success

> Those who destroy an enemy state, while preserving their strength and prosperity for later generations, are true kings. Those who destroy an enemy state only to let their strength and prosperity accrue to the neighboring states are doomed to disaster and downfall.

In this final lesson, Master Guan warns that if you set out to destroy an enemy you must not destroy yourself in the process. Perhaps the most famous example of this is when Croesus was preparing to attack the Kingdom of Persia and asked the Oracle of Delphi for advice. The Oracle said, "If King Croesus crosses the Halys River a great empire will be destroyed." Croesus instigated the war but ironically destroyed his own kingdom instead of ending the Persian Empire.

The purpose of military conquest should be to insure your security and enable your people to thereafter experience peace and prosperity. If you exhaust so much of your strength that there is nothing to pass on to subsequent generations and if too many innocents are destroyed in this quest then you will simply have wasted yourself and actually failed in your objective.

Those who destroy an enemy state only to let their strength and prosperity accrue to their neighbors may be setting the stage for future conflict by empowering future contenders. This is like the folly of the 2020-21 American retreat from Afghanistan where it left behind billions in weapons that were sold throughout the region. By undertaking military moves to weaken others, as the United States is found of doing, you might be weakening yourself and strengthening your future adversaries.

Hence the supreme leader always asks himself, "What am I leaving for the future generation?"[62] If the answer is very little then his actions are in the wrong.

A king or national protector militarily intervenes to create peace for his state, to restore order from chaos, and to save states from the brink of destruction. He doesn't set out to annihilate people or drive states into extinction. When a world leader intervenes to dethrone a despot he acts to remove the corrupt leadership rather than crush the populace. He harbors no hidden scheme of imperialism or expansionism in his behavior. The only consideration is a good outcome for people who have suffered as the victims of a bad leader, who is the actual enemy.

Since it is the innocent who suffer when a leader makes mistakes you must always be careful what you do when you hold great political power. To correct what is wrong and rid the world of what harms the people, to strike down the aggressively despotic, to harmonize what is uneven, to secure the imperiled, to renew the people and pave the way for future generations of stability and prosperity—these are the objectives of a supreme leader.

Those who accomplish these goals while broadcasting the cause of righteousness and virtue, who can embrace all the world's people as if they were a single family, who can consider their subjects as friends and respect them as if they were guests, who refrain from global ambitions that will harm others and national ambitions that will imperil their own state, who serve the best interests of the world—such can be great kings. As the Sanskrit *Art of Wealth* says:

---

[62] Great leaders usually adopt the daily practice of "mindfulness," introspection, or "contemplating one's self" to stay in touch with reality and their own conduct. This is practicing a continuous inspection of one's mind and behavior to prevent errors in thinking and conduct as well as insuring wisdom and honesty when determining policies for the nation. Through this means of self-cultivation, and by listening to the advice and criticism of others, they become able to correct their own prejudices and errors before they can materialize into trouble for the state. The condition of the state starts with its roots, which are a leader's actions and behavior, and these originate within the political leader's mind. Only by correcting the root source behind a nation's woes, which is the leader's thoughts and prejudices, can one keep a country from destruction.

> One who is protector of the helpless, refuge for refugees, guide to the lost, haven for the fearful, supporter of the disenfranchised, friend, kinsman, patron, resort, benefactor, boon, teacher, father, mother and brother to the world, that one is a leader.[63]

"Will my actions benefit the country's future generations or destroy my own nation?" This is the question the sage king always asks himself, and the yardstick by which he will be measured in history. Inspect yourself before you set out to gain political power to see if you make the grade. Are you trying to help yourself or others?

Fortune or disaster always starts within your own mind and then manifests because of your subsequent actions and behavior. If you examine your thoughts and intentions you will know whether you are self-centered or people-centered, and whether you are trying to help others or yourself while feeding your own ego. Everything comes down to virtuous intent rather than self-aggrandizement and if this one component is lacking then the rest should not be considered.

Leadership always starts with personal cultivation. You must first master yourself before you are qualified to exercise power over others.

---

[63] *The Art of Wealth*, trans. by Thomas Cleary, (Deerfield Beach, Florida: Health Communications, 1998), p. 141.

CHAPTER 4
ECONOMIC WARFARE TO CONQUER
WITHOUT FIGHTING

There are many other important lessons we can learn from the *Guanzi* as regards leadership, political power and forging a state into a superpower. One particularly useful lesson is Master Guan's method of combining monetary and fiscal policy to make a state into a formidable economic power, and to even conquer enemies through economic warfare. Before we delve into this topic, however, there is still an important issue to discuss, which is a leader's need to be able to understand human nature.

Every leader needs to develop a deep understanding of human psychology and behavior. For instance, in order to use men or serve men you need to be able to understand their hearts. In particular, in order to lead others to a better future you need to understand what they want and be willing to help enrich their lives. This is why, when asked about the "good ruler," Confucius simply replied, "He loves men." When further asked about the "wise ruler" Confucius replied, "He knows men."

That Guan Zhong was himself adept at this skill is evidenced in the fact that he remained at the helm of political power until his death. There is one particular story that shows his great wisdom in understanding human nature.

At one time, Duke Huan asked Guan Zhong how to relieve the state's population of the burden of usury which the people had entered into so as to support the government during a time of war. Guan Zhong therefore ordered that in each county, the homes of the money lenders be granted the favorable distinction of white washed doors and high lintels. He then told the duke to send envoys presenting jade disks and greetings to the money lenders, thanking them for their services which were, of course, prompted only by material gain. His greeting to the money lenders went,

> The *Book of Poetry* says that "Kind-hearted gentlemen are like fathers to the people." Because you extended loans to the poor so that they could meet the urgent needs of the government, they were able to continue cultivating their fields in accordance with the season without any disturbance to the state. All this was made possible by your generosity. Therefore, these gifts are given to you as a token of thanks and as a compliment, for you are as good as fathers to the people.

As a result of receiving such recognition for what was essentially purely a business transaction absent of any virtuous intent, many of the money lenders destroyed their debt contracts and distributed their own wealth to the people thus relieving the population of their burden. Understanding human nature, Guan Zhong reasoned that the money lenders would do this (which was his strategy all along) so as to live up to the good name they had just been awarded. He knew that people who have wealth tend to value what they cannot buy and so he proposed this plan to take advantage of their inner desires. In this manner, Guan Zhong used his wisdom and understanding of human nature to achieve the duke's objectives and accomplish what could not be accomplished in any other way.

No one will deny that a leader needs to understand the psychology of his own people and that of other nations in order to gain power and make his country more prosperous. Being able to understand the human mind is an important component to becoming a leader and gaining power. The cycles of history repeat themselves simply because human nature doesn't change. It remains basically the same whether we speak of ancient times or the new. It's only after you understand what people like and dislike that you can speak to them in an attractive way, talking about what interests them and addressing their inclinations to motivate them along your avenues of concern.

In addition to psychology it is also extremely necessary for pacifistic world leaders to be on familiar terms with the subject of warfare. Master Guan always preferred diplomacy to fighting, but he also did not shrink from a fight. He fully recognized the *Tao Te Ching* in that

> One who excels in conquering the enemy does not engage in
> battle with them;
> One who excels in employing men acts deferentially to them.

Hence diplomacy was usually his first and foremost course of action.

When diplomacy failed then instead of using military measures Master Guan came up with brilliant strategies for taking over foreign states through economic tactics rather than through warfare. Master Guan quite believed that those skilled at warfare never fight at all. Warfare is unpredictable and will cause you the loss of lives and wealth. As Sun Tzu also said, "To win without fighting is best of all."

On the general topic of war, Master Guan felt that no true citizen could remain indifferent to his country's danger, and so in times of peace a nation still had to be prepared for the need to fight.[64] Switzerland, although famous for its neutrality, therefore keeps its citizenry well trained for military action and those policies help to unite the nation. A nation should always accumulate wealth during peacetime because those resources will serve the nation during warfare. Though Master Guan preferred peaceful strategies over military measures, he always insisted that a nation be prepared for war anyway as it could happen to anyone.

This is a quite different attitude than that of the "love all the way round" crowd. While nations do not have to mimic Sparta or Israel in military preparedness they must still be ready for the eventuality of conflict. While your nation might be surrounded by friendly neighbors and might refrain from initiating antagonistic moves there is no certainty that your neighbors will not attack you one day. If that happens, no set of hastily assembled troops will be able to match the trained forces of an invader. Since there's no such thing as a world court or policeman to prevent invasions a country should always be prepared for war.

Our modern world has seen the development of long range ballistic missiles, cruise missiles, long range bombers and drone warfare that have essentially made everyone a close neighbor within reach of ready destruction. What leader—of yesterday, today or the future—can bear such a risk with nonchalance when a single day's attack can destroy an entire generation's worth of labor.

Master Guan felt that a nation had to have it's own house in order and nurture its strength before it launched any military campaigns. A nation must always be strong enough to initiate a war or defend itself against invaders. A strong economy is therefore necessary for military defense, not to mention offense. Allies are not always dependable so a nation's economy is its only dependable source of support.

In Guan Zhong's day, sound economics started with protecting the

---

[64] Ancient Rome kept its peace and lessened the threat of foreign invasion by making sure its frontier armies were well-prepared, which is why the barbarian tribes surrounding Rome would say, "The Romans prepare for peace by acting as if they were at war." As activities in the Middle East have recently shown, the need for war preparations are as necessary today as they were hundreds of years ago.

farmer because farmers produced grain, which fed troops, who in turn fought to protect the country. Master Guan therefore organized his country's economics, and the conscription of armed forces, to protect the root source of the nation's wealth.

In terms of tactical strategy, Master Guan was like many other great generals who would use all sorts of ruses to win a war. However, he emphatically disapproved of any forms of trickery and treachery once an enemy was defeated. He spent an incredible sum on foreign relations, which meant paying off (bribing) foreign officials when he could and used diplomacy as a weapon in order to advance Qi's interests. This strategy may seem costly but is always superior to risking destruction through military conflict.

On this note, it is useful to consider the modern example of Japan. Because it is a small island nation dependent upon other countries for raw materials and markets, and because it lacks strong military capabilities, Japan's development strategy has followed Master Guan's thinking in many ways. In the 1980s, wherever possible the Japanese effectively bought off[65] the influential opinions of many foreign officials and key decision makers while staying invisible in the background. It is futile to fault a nation for acting in its own strategic interest, but you can fault those nations who remain ignorant and blind to such maneuvers. All countries act in their own self-interest and will use blackmail, bribery, political pressure, economic pressure, military means and other methods to get their way.

Master Guan's special genius in foreign policy, as we previously noted, was to use economic means, rather than military undertakings, to increase his country's prosperity and expand its sphere of influence. As a result, he often manipulated commodity prices instead of using warfare to destroy other states. In a massive intelligence operation the United States used this same strategy by suppressing oil prices to bankrupt the Soviet Union.

By intentionally making the price of certain commodities much lower in his state than abroad, Master Guan could entice the populations of neighboring states to migrate to Qi as new subjects, thus strengthening his state. When a foreign state lost its farmers and their agricultural production to Qi it was only a matter of time before such a state would fall under Qi's influence.

As an illustration, Master Guan might choose to initially purchase all the military weapons produced by a particular state at high prices, which he would then resell to other territories at cheaper prices. Upon hearing of his

---

[65] As Pat Choate (*Agents of Influence: How Japan's Lobbyists Manipulate America*, Simon & Schuster, 1990) and others have shown. I remember reading this book in the 90s and saw many in the press and in power ignoring his observations because they were receiving financial benefits from the Japanese.

purchases, competing nations would also buy weapons from that state to insure their supply. The end result was that citizens of this weapons-producing state, the actual target of his ambitions, would gradually abandon their regular farming because they could make much more money by growing wood and producing weapons.[66] In other words, they shifted their economic activities away from farming to weapons production because it was far more profitable.

Continuing with his strategy, Master Guan would next manipulate Qi's grain market so that prices rose, with the result that surrounding territories would send most of their grain into Qi in order to capture the higher prices. Next, he would suddenly close off his frontiers so that the arms-producing state, possessing now neither arms (since it was drained of weapons) nor grain (since the farmers had devoted themselves to weapons-making rather than farming), would face possible invasion from its neighbors and have no choice but to surrender unconditionally to Qi. This is how he would practice economic warfare so that he could win a war without fighting. An ignorant country whose people can be manipulated into becoming more vulnerable while thinking they were making a profit is a perfect candidate for conquest.

The United States used a similar ploy in lowering the price of gold and oil to drive Russia into bankruptcy and collapse the Soviet Union.

Under the guise of "development advice," a number of African states were tricked in abandoning their own native produce, which was perfectly suited to the local soil, climate and culture, in favor of cash crops such as corn that were better suited for the West. These states thought they would increase their agricultural output and gain export earnings by doing so. However, these imported crops greatly depleted the African soils and required extensive amounts of expensive fertilizer imports. These crops depleted the native soils and drove the nations into debt so that they became the pawns of foreign powers. African states who thought they would gain actually became losers by following the best foreign development advice, which was meant to cheat them and make them fall under the control of the West.

In his book, *Confessions of an Economic Hitman,* John Perkins described

---

[66] Do people actually act in this way? Indeed so, for as seen in the bull market euphoria of the 1980s, many individuals mistakenly abandoned their regular jobs in hopes of living off rising stock market prices. This naiveté produced particularly painful unemployment when the markets eventually crashed. Similarly, when any manufactured item becomes "hot" in the marketplace, companies commonly reallocate their resources so as to produce more of this product until a glut eventually emerges, driving prices down to hurt all the players. Such boom-bust cycles are the way of the world, and it was Guan Zhong's genius to make use of this for geopolitical designs.

the practice of highly paid professionals inflating financial projections to justify the need for massive infrastructure projects in developing countries. Once those countries bought into approving the oversized projects, which required massive international loans that were intentionally too big to repay, they ended up with foreign debts hanging over their heads. The international bankers and foreign governments would then twist the arms of those countries to gain political influence in those nations or in portions of their economies.

China is using the same strategy today to fund infrastructure projects in willing nations for its Belt and Road initiative. China provides loans, countries use its construction companies and labor to build the projects, and China hopes to sell goods and services along those transportation lines to fuel its economy for several decades into the future. By creating an inland railway and road transportation system that is out of reach of the range of western bomber strikes China hopes to reduce its dependence on maritime trade with its susceptibility to western naval blockades, and thus is making moves to bypass the ability of western nations to thwart its trade routes. It is only a matter of time to see whether a flurry of nations default on the loans.

There is also the well-known practice of large Japanese trading houses (*sogo sosha*) which send representatives into underdeveloped countries to pay an unusually high price for a selected commodity. In so doing, they play a form of "worshipping the skull" by advertising that there is a lot of money to be made from farming the commodity in question. Soon farmers (are frantically trying to increase their crop yields with the result that in the next harvest season there is a tremendous glut of supply in the market. Naturally, the commodity's price then plummets and cheap purchasing becomes the norm. Having achieved their intended effect, and ruined many people in the process, the Japanese then stock up on the commodity at bargain basement prices.

At one time Duke Huan wished to conquer the neighboring states of Lu and Liang, and asked Guan Zhong how to do so. Guan Zhong replied that the people within these two states produced a certain type of coarse fabric. If the duke wore garments made from this fabric and ordered his ministers to also do so then the people would imitate their lead and choose to wear it as well.[67] Then, all Duke Huan had to do is forbid his own people

---

[67] This principle holds good in terms of morality as well, for as the famous *Neeti Scripture* (an ancient Indian code of regal conduct written more than 2300 years ago) relates: "The righteous king induces his subjects to righteousness, the sinful to sinfulness, the equitable to equitableness. Because the subjects follow the king, as is the king so are the people." Basically, people follow the lead of their ruler; whatever his personal conduct, so does the conduct of the populace become. People follow what their leaders do, not what they say. The leader is like the wind and people like

from manufacturing the material so that the people had to import it entirely from abroad. He surmised that Lu and Liang might see tremendous profit possibilities from producing the material, and then they would abandon their farming and devote all their efforts to manufacturing the fabric.

Duke Huan immediately had a robe made and started to put the plan into effect. In the meantime, Guan Zhong told the merchants of Lu and Liang that if they made this fabric for Qi they would make so much money from Qi that they wouldn't even have to tax their own people! Everyone using economic warfare is always trying to trick their opponents into following bad advice. With such a good story, the sovereigns of Lu and Liang naturally ordered their people to manufacture the fabric.

Several months later, after spies reported that the rural areas of Lu and Liang were frantically busy with commerce involving this material, Guan Zhong told the duke to close Qi's frontiers and sever diplomatic relations with the two states. Ten months later, the people of Lu and Liang were starving and had nothing with which to pay their taxes. Even though they had again started to grow grain, a harvest could not be expected for at least several months and hence the price of food in Lu and Liang rose to one hundred times the price found in Qi. Within two years, over sixty percent of the people from these states migrated into Qi and within three years, the sovereigns of Lu and Liang petitioned their surrender. This was a strategy designed with the knowledge of human behavior and appropriate timing.

In another instance, Qi once offered to buy silver fox fur from the state of Tai. Fox fur is a very scarce commodity but the people, expecting to sell it at an extremely high price, widely abandoned their occupations to track silver fox in the forests and mountains. When a barbarian tribe heard of this they attacked the state of Tai prompting the king of Tai to ask for his state to be attached to Qi. Duke Huan had not spent a single coin or even sent any diplomatic mission to Tai and yet he was able to annex the state!

As a final example, the state of Qi once offered to buy deer at high prices from the powerful state of Chu. In this particular instance, Chu felt it was outsmarting Qi because deer are injurious to farmland. Qi was therefore effectively offering to drain its valuable monetary resources and buy harmful items that Chu wished to eliminate. How could this be anything other than a blessing from Heaven? However, the plan worked just as before.

While Chu did become momentarily richer, its people eventually abandoned their farming to hunt deer because it was a more profitable occupation. When Qi finally closed its borders and severed diplomatic relations with Chu, the price of grain in Chu rose so high that almost half of

---

grass, so that when the wind blows, the grass bends.

Chu's population migrated to Qi, and it also surrendered to Qi.

Can strategies such as these actually bring countries to their knees in the modern world? To answer this you need only consider the fact that many smaller nations are still commodity-based economies. Unless they join worldwide cartels that control the prices of their major products—whether they be bananas, coffee, oil or tin—the economies of such countries will be highly subject to the manipulation of foreign parties.

Economic blackmail, such as threats to impose tariffs and import quotas on a nation's products, can indeed influence a state and such non-military strategies have often destroyed nations. One need only look at the example of the Rothschild banking brothers to see that even private individuals can hurt governments in this fashion. To punish the Spanish government for abrogating the terms of their Rothschild loans, the banking brothers once engineered a heavy fall in the prices of Spanish government stock to dissuade others from also reneging on their debts.

The possibility of excessive foreign influence on your nation from holding your nation's debt is a reason that U.S. Treasury officials become concerned when foreign nationals purchase large blocks of U.S. interest rate instruments. Large foreign purchases definitely make the U.S., and any other country in similar circumstances, more vulnerable to external market manipulations if foreign purchasers were ever to act in consort. The Treasury, for instance, fears the reality of Project (Operation) Sandman which is said to be a plan by foreign governments to simultaneously sell their large Treasury bond holdings and crash the U.S. dollar by working en masse.

This potential for an outsider's influence on a country is one of the reasons why many countries impose restrictions to limit foreign participation in their real estate and capital markets. National protectors consciously recognize such strategic concerns even though most government officials in host nations remain largely ignorant of on-going geopolitical designs against them, or are simply not in positions to do anything about them. Front line company executives, whose firms are subject to the risks of supplier blackmail or bankruptcy, are never blind to the risks of outside control. Therefore they always try to find multiple suppliers for their necessary raw materials.

While we might criticize all these schemes of economic blackmail, even Confucius approved of such tricky maneuvers when they were designed to save a nation. For instance, at one time Confucius was residing in his native state of Lu when he heard that the Qi army was preparing to invade his state. Gathering his disciples together, Confucius announced, "Lu is the site of our ancestral temples and tombs, the land of our parents and relatives. How can we sit here and watch it fall victim to foreign invaders while the souls of our ancestors cry out from under the earth? It would be a blessing for Lu if one of you students could dissuade Qi from action."

First Zi Li, the bravest of Confucius' disciples, volunteered to take action, but instead Confucius approved only when Zi Gong, who was known for his clever thinking and elegant words, volunteered to go to Qi and talk to its chief minister, Tian Chang.

Arriving in Qi, Zi Gong had an audience with Tian Chang where he said, "The state of Lu [where Confucius resides] is not easy prey, for it is small, has poorly fortified cities and is weak with incompetent ministers, disorderly soldiers, disloyal citizens and an ignorant sovereign. It will therefore be difficult to overcome. It is better to attack the state of Wu instead, for it is large, well fortified, has wise ministers and also a powerful army. Therefore it will be much easier to capture."

On hearing this obviously backwards statement, Tian Chang raged, "What kind of nonsense is this? Why are you mocking me?" Zi Gong, however, was not startled by this response at all. Rather, having captured Tian Chang's attention, he silently beckoned that the minister's attendants be dismissed so that he could discuss matters in private.

"Right now you are just the prime minister but want to become the ruler of Qi," he whispered, "and so will face enemies within rather than outside the state. If you attack the powerful state of Wu, you can turn it into a weapon against your internal enemies. If the Qi army wins against Wu, the victory will be yours whereas if it fails, many generals and ministers will die, offering you just the opportunity you need to take control of Qi. If you attack the weak state of Lu, however, you are assured to win and all the other generals and ministers will share your glory thereby strengthening their positions. That will make it even more difficult for you to carry out your plan of seizing power."

Tian Chang was shocked on hearing Zi Gong's words for he realized that his uninvited guest was right. His major reason for attacking Lu was that he only wished to further consolidate his reputation and political power before usurping Qi's throne. Tian Chang therefore replied, "What you say makes sense, but it is too late to withdraw now because our troops are stationed at the border. It would be difficult to turn them towards Wu. Such a move would damage the soldiers' morale."

"That's easy to take care of," replied Zi Gong. "I'll go to the state of Wu and talk its king into attacking Qi. Then your men will be justified in attacking Wu instead of Lu." Tian Chang readily agreed to this plan.

Zi Gong then traveled to the state of Wu, where he met its king, Fu Chai. "The states of Wu and Lu jointly attacked Qi years ago. Now Qi is planning revenge and has sent its troops to attack Lu, after which it will try to conquer your state. Why don't you send your army to save Lu?" he asked. "If you have Lu as an ally, together you can defeat Qi with a preemptive strike. Then with Lu as your ally, you can use the forces of both states to realize your military ambitions. You can then challenge the state of Jin for control of China's central plains."

King Fu Chai replied, "I have been thinking of punishing Qi for quite some time, but I hear the state of Yue has been building up its power because it is plotting against me. It would be better for me to conquer Yue first, who wants to attack me, before I attack Qi."

"I understand what you're saying, but there is a flaw with your plan," countered Zi Gong. "If you attack the state of Yue first, Lu your ally will have already been swallowed up by Qi. Actually, the king of Yue should not worry you anyway. I can persuade him to lead troops to help you in attacking Qi." Fu Chai agreed to this suggestion, after which Zi Gong headed towards Yue to convince the king as planned.

After arriving in Yue, Gou Jian, the king of Yue went personally to see him. "I know you have taken the trouble to come a long way to visit our state so you must have some important suggestions for me."

"Actually," replied Zi Gong, "I've come to express my condolences. King Fu Chai of Wu suspects you are plotting against his country, and wants to destroy Yue before attacking the state of Qi. If that happens, your state will have no chance of survival."

"It is indeed fortunate for us to hear of your timely message.

What do you suggest we do?" replied Yue's sovereign.

"Fu Chai of Wu is arrogant and quick tempered. But if you present Wu with rich gifts and humbly offer to help him in subduing Qi he will be pleased and his anger will be mollified. Consider that if he then fails in his attack upon Qi you'll benefit because Wu's forces will have then become weakened. If he wins success, however, he'll next turn his ambitions toward the central plains and contend with the state of Jin. That's when you can use Wu's laxity, since you helped it attack Qi, to plan any attack of your own that you're considering. If you help Wu you will benefit regardless of the outcome."

"You must have been sent here by Heaven!" replied Yue's sovereign, "We'll follow your plan just as you've suggested."

Zi Gong then returned to the state of Wu and reported back to Fu Chai. "The plan is working just as we expected. The king of Yue was frightened when he heard your words, and is sending gifts to make any amends. He promised that he will send troops to help you invade Qi, but it would be improper for you to employ King Gou Jian himself in the expedition because a king never commands another king. However, you should use the troops he will send." Fu Chai received the gifts and use of troops, just as Zi Gong had reported, and so led an attack upon the state of Qi, which was Lu's original antagonist.

However, Zi Gong now realized that Wu might become a danger to the state of Lu anyway, because Wu might turn upon Lu for its prey if it indeed succeeded in defeating Qi. He therefore immediately proceeded to the state of Jin with the idea to use Jin as a counterbalancing influence to discourage Wu from swallowing Lu. After meeting Duke Ding of Jin, Zi Gong said, "The state of Wu and Qi are preparing for war. If Wu suffers defeat, the state of Yue will attack it from the rear. However if Wu defeats Qi, Wu will challenge your state for dominance of China's central plains. You should get prepared for this eventuality by exercising the troops and by preparing weapons and provisions." Seeing the wisdom in this, without knowing of Zi Gong's grand design, Duke Ding did exactly as advised.

In a short time, Zi Gong's complicated interrelated strategy for saving Lu worked to perfection. King Fu Chai led the combined forces of Wu, Lu and Yue to crush Qi, after which he arranged for Lu and Qi to swear an oath of brotherhood so that his neighboring states would have peaceful relations. Therefore, as Confucius had wished, the state of Lu was saved from invasion by Qi. With victory behind him, King Fu Chai's next target became the state of Jin just as Zi Gong had expected. But Jin was well prepared due to Zi Gong's warning and defeated Fu Chai in battle. Not being able to immediately realize his ambition, Fu Chai kept Lu as an ally.

By entangling these four states in interlocked strategies, Zi Gong successfully saved the state of Lu. Despite his use of deception and intrigue, his plan was something that Confucius could approve in order to save his own nation. When you need not do the fighting yourself, but can "make use of a borrowed knife" by getting another to do the work for you, such a strategy is worthy of Guan Tzu. For instance, the British and Israelis always attempt to maneuver the United States into fighting their wars.

If you truly want political power, geopolitical power and also want to protect your nation's people then you must be aware of the designs that other nations might try to employ against you. Everyone is involved in trickery with the intent to fool everyone else to their own benefit. This is the way that geopolitics is played.

On the other hand, to know economic warfare yourself and use it to save your own nation is indeed a great blessing for your nation. At the same time, continually depending upon craftiness and trickery will ultimately lead to a nation's downfall and ruin because the more it is used the more it will spread into all facets of the government and then society. Therefore, you must think of a hundred years of future history before you put such strategies into play because consequences can last that long and nations that you attack have memories that hold grudges for that period and longer.

CHAPTER 5
ELIMINATE YOUR INTERNAL ENEMIES

By following Guan Zhong's advice Duke Huan of Qi was eventually able to extend his influence to over all of China. He became the most important person in the land and the first of China's five historic "Lord Protectors." Despite the fact that Guan Zhong had originally tried to kill the duke when he was racing for the throne, a strong bond had grown between the two to the extent that the duke now addressed Guan Zhong as if he were a member of his own family.

After many years, Guan Zhong eventually fell gravely ill and the duke went to visit him. "Grandfather Guan," he said, "your illness has become very serious and I am afraid you may not recover. If you don't make it through, what final advice do you have for me and the state?"

"I would have brought this very matter up myself," replied Guan Zhong, "except that I feel you will not be able to bring yourself to do what I suggest."

The duke replied, "Grandfather Guan, if you were to instruct me to go east, I would go east. If you were to tell me to go west, I would go west. Whatever you were to command, how could I possibly dare to ignore your advice?"

Guan Zhong then arranged his clothing and straightened himself to sit upright in bed. "Good, good," he said. "As my final time is coming, I have only this last and most important advice to give you. The four men, Yi Ya (chief cook), Shu Diao (chief of the eunuchs), Tang Wu (court sorcerer) and Gongzi Kaifang (a leading courier) should all be dismissed and sent far away, or else you will be in trouble. The reasons are as follows.

"When the Master of the kitchen Yi Ya heard that you had never tasted human flesh, he killed his own son and steamed his head so that he could present it to you for tasting. It is certainly human nature to love one's own children dearly. If Yi Ya doesn't even love his own son that he was able to do this, butchering him like an animal, do you think he really has any genuine love and concern for you? This man, by his actions, shows that fame and power are more important than the life of his own son. Can you therefore expect him to be trustworthy and faithful? You must get rid of this man.

"Your Grace also takes great pleasure in his harem and is very jealous of his women. Now your retainer Shu Diao had himself castrated so that he could personally serve you by managing your harem. It is human nature to love one's own body and self before that of others, but if Shu Diao does not treasure even his own body enough that he was able to do this then how can he genuinely treasure yours? Since this man mutilated his own body to get into your trust you must realize that he will do almost anything to obtain his goals. You must get rid of him as well.

"Tang Wu can provide supernatural answers to all sorts of your questions, but the matters he touches upon, such as illness or life and death, are matters decided by Heaven. No one can guarantee our fortunes for us so how can this man and his advice help you when he has no control over these matters himself? There is too much danger in listening to this man's advice and granting him extra privileges. Not only will his prognostications cause you to miss the big picture because you become transfixed with minor details, but it will lead him to become arrogant and engage in wicked ambitions to your detriment. You should not use Tang Wu anymore, but should also dismiss him because he represents a danger to you.

"Finally, Gongzi Kaifang has served Your Grace for over fifteen years, but in all that time he has not once returned to see his family even though the distance is but a few days travel. Consider it carefully, Your Grace. If Gongzi Kaifang does not love his parents and family enough to visit them, not even leaving court when his father died, can he really have any genuine feelings for you? You must remember that in order to serve you Gongzi Kaifang left his position as the heir apparent to a thousand chariots. The reason he did this is not out of any love for you but because what he really wants is more than a thousand chariots. It's normal for people to love their own family over that of another. If Gongzi Kaifang doesn't love his own family where he never visited them in all this time, do you then really think he has any genuine love for you? You must definitely dismiss him as well.

"These four men have been held in check while I have been alive, but after I am gone you will be in danger since there will be no one to restrain them. Therefore, you must dismiss them.

"I have often heard that individuals who perform their duties under false pretenses, which they try to conceal, will not last long and that people who do not lead good lives will also come to no good end. So this is my final advice. Please mark my words and heed my warning."

"I understand all that you say," replied the duke, "and promise to carry out your recommendations."

Subsequently, Guan Zhong's illness turned for the worse and he finally passed away. After his death, the duke remembered his advice, summoned the four men and eliminated their positions. However, after dismissing the divination master, Tang Wu, the duke's mind became unsettled. After dismissing the master of the kitchen, Yi Ya, the duke lost his taste for food. After dismissing the chief of the eunuchs, Shu Diao, there was confusion and disorder in the inner palace, and after dismissing Gongzi Kaifang the absence of beguiling words and sweet phrases made court life seem unbearable to the duke. Soon the duke was doubting Guan Zhong's advice and had all four men reinstated.

After a year's return to court, the duke became ill and Tang Wu used the opportunity to create a rumor that the duke would soon die. The four men then launched a coup and imprisoned the duke in an isolated room cut off from communication with the outside world. Eventually one of his wives was able to smuggle herself into the room in order to see him.

"I'm starving," the duke said, "but they won't let me have anything to eat or drink. What's the reason for this?"

His wife replied, "The four men Yi Ya, Shu Diao, Tang Wu and Gongzi Kaifang have been using your name issuing orders to the other officials for some time to interfere with the state's affairs for their own benefit. Now they are dividing the state of Qi and parceling out the land. The roads have been blocked for some time now and so food is unobtainable."

"Alas that it should come to this," sighed the duke. "The words of the sage were so right. If I had not known this I could die in peace, but now that I am dying and know how can I face Grandfather Guan in the afterlife?"

Thereafter, the duke picked up a white scarf, wrapped it around his head and passed away. It was only after people saw the maggots around the door, several days later, that they realized the duke had died. They covered his body over with a leaf from the south gate of the city and placed him in a coffin, but months later he still had not been buried.

Because he did not employ worthy people toward the end of his life the duke, who had reigned supreme, was dead for eleven days until the worms around his door announced his death.

Let the many years of Duke Huan's spectacular success, followed by his inglorious end, serve as a warning to all that they should heed the wise advice of sagacious men.

CHAPTER 6
LIFE PURPOSE AND POLITICAL POWER

What's the ultimate purpose behind all this discussion on gaining political power and ultimate political supremacy? What's the ultimate goal? If you are a political leader or aspiring leader the goal should involve your views on the meaning of life and your life purpose.

Confucius said that part of the "Great Learning" in becoming a true human being is demonstrating your love for people by helping them in all ways. How do you show your love for the people? By cultivating your behavior, by taking responsibility for your own actions, and by taking definite steps to bring peace, prosperity and harmony to the world.

You "intervene in the affairs of the world" for the compassionate aim of trying to save the world. This, in fact, is the goal of the Confucian sage king, Taoist sage, Buddhist Bodhisattva, Indian Avatar, and the Christian savior. This is the purpose of the enlightened leader. A "One World Government" is not the way, however, because the Roman Empire with its *many civil wars* proved that it would never produce peace. Communism, Marxism and Socialism have never produced prosperity or peace either and have been responsible for hundreds of millions of deaths just like religious wars. Most theocratic states have failed miserably too. Fascism has never released a wide variety of creative interests in a society, lead to a high measure of cultural advance, or maximized human happiness while enabling the maximum of personal freedoms compatible with the freedom of others in society.

At the cost of being repetitive, Lee Kuan Yew said that a leader should try to produce the maximum happiness and well-being for the maximum number of people. If we were to turn this into various goals that might serve as affirmations for leaders we might create a list such as the following:

THE ART OF POLITICAL POWER

- Enrich the lives of people everywhere with your efforts
- Maximize the benefit to the people and protect their interests
- Establish conditions that improve standards of living for the majority of people
- Try to create value for the people
- End their sufferings and afflictions
- Bolster the positive DNA of the essence of the culture
- Create a brighter future for the people. Provide them a path to a happier life
- Establish policies and actions that serve their happiness and well-being
- Establish conditions that inculcate within society the social qualities that will be useful in the building up of society
- Give the people a sense of purpose and participation
- Nurture excellence and encourage the average to improve
- Make the people great in order to make the nation capable of greatness
- Release a wide variety of creative interests
- Enable and protect the maximum of personal freedoms compatible with the freedom of others in society
- Protect and support the poor, sick, elderly, oppressed, defenseless, etc.
- Produce order and justice in the relationship between person-person and person-state
- Restrain evil in society and prevent the powerful, wealthy, government, elites, and monopolies from oppressing or exploiting the common man
- Shape the evolutionary process to optimize human prosperity, human happiness, material and cultural progress, and self-actualization
- Magnify the people's spirit
- Bring out the country's magnificence
- Advance the spirit of the age to a new point of excellence

Your country cannot gain maximum geopolitical power unless it also becomes rich. In terms of the formula for increasing national prosperity the best book I ever found on grand strategies for national economies is *How Rich Countries Got Rich … and Why Poor Countries Stay Poor* (Erik Reinert). The foundation of prosperity for a country is economics. Each country has other nation states that are its competitors that desire to either destroy it, conquer it so that it can be used for resources or labor, or suppress it by thwarting its development so it does not end up stealing trade. The general strategy used by others is to keep a developing country at a lower strata by any means possible, including by giving bad advice.

The key to economic development is that countries need to diversify away from commodity production to industrial production. They need to switch from simple raw material, timber, mining, fishing and agricultural production (sectors with diminishing returns to scale) and move to sectors with *increasing returns to scale* such as manufacturing, technology, and services. A move to high quantity manufacturing insures that profit margins will increase over time as volume builds while the productivity experience curve cuts costs and time due to your experience. Manufacturing also commonly leads to innovation and thus other new products that can become new sources of revenue. Most of all, it can lead to brand name positioning that can claim a premium markup over competitor products.

Commodity production, on the other hand, offers no chance for premium pricing or monopolistic profits whereas a brand name product does. All raw material production is eventually subject to diminishing returns over time and thus higher costs as time goes on. For instance, it costs more money per bushel to grow a larger quantity of corn after the most fertile land is used up, and it is more expensive to mine additional copper as the richest ore deposits become depleted. Raw material production is often used to gain initial revenues and capital for a nation but is not a viable strategy over the long run due to decreasing margins, increasing competition, and the inability to differentiate yourself and secure premium pricing. Countries that base their economies upon commodities rather than manufacturing eventually go bankrupt as their costs rise, competitors grow, and they run out of those raw materials.

Most western countries became rich by diversifying their trade and industrial base, and by having their governments grant monopolies and use tariffs to prevent cheaper imports from hurting domestic manufacturing startups. The first rich countries (such as England, Holland and Italy) developed their economies via diversification and by protecting their efforts at industrialization (see Appendix 2 in my book *Bankism*). These countries benefitted more from selling finished products than from selling raw materials (commodities), and so they employed tariffs and sanctioned monopolies to protect their early industries that produced finished goods.

This is the general path to national riches: produce value-added goods (rather than commodities and raw materials) and protect those industries as they grow. Those nations which have employed this strategy have always become wealthy. For instance, America was also once highly protective of its manufacturers for over two centuries (despite Britain urging it not to do so in order to prevent competition) and only gradually opened its industries to international trade *after* those industries were ready to bear the heat of world competition.

Aside from the principle of benefitting from declining costs over time due to the experience curve and therefore making a greater profit margin over time, the second half of the equation is *selling more of whatever you are making*. To do so you need better marketing and more (larger) markets in order to bring in greater and greater sums of money. In fact, many wars have been fought to force countries to open their markets to a nation's imports. Countries will become wealthier if they gain more markets for their products. It is all a game of taking money from others and bringing it back to you.

For a country to become rich it must sell products in its domestic market, gain enough experience to cut costs and improve quality, and then eventually progress to international trade. You cannot become rich if you do not offer something to the rest of the world (so that their money flows into your country) but what you offer should be manufacturing products or other goods/services that embody increasing returns to scale efficiencies. Only increasing returns to scale types of activity allow you to enjoy higher profit margins and make more money as you produce more. Financial services, for instance, fall into this category.

If you enjoy increasing profit margins while simultaneously enjoying an increasing demand for your products/services then the multiplication of these two factors together produces wealth. This is the secret behind nations becoming wealthy. If demand for your products grows while your margins improve this is the formula for wealth that most countries have used to get rich.

The focus on becoming wealthy should involve a push to develop domestic manufacturing industries (or other high margin businesses), which is the means that the rich countries of the West themselves all used to become wealthy. It is not that they grew rich through manufacturing. They grew rich through "increasing returns to scale" activities while also seeing an increasing demand for those products.

Rich countries tend to specialize in man-made (manufacturing) comparative advantages while poor countries usually focus on nature-made (raw material) products as in farming, fisheries, timber or mining. Man-made advantages lead to increasing returns to scale while commodity production (nature-made) leads to decreasing returns to scale, and it impoverishes a nation over the long-run as natural resources are gradually used up or become harder and more expensive to procure. Furthermore, one can also *brand* a manufactured product by making it unique and different than every other competitor to thereby help increase its demand, but it is nearly impossible to compete on anything but price when your output is a standard commodity produced by many other world producers.

Countries seeking the road of wealth should also think of developing economic complexity (highly differentiated manufacturing production) through diversification rather than solely relying on commodity exports such as oil, coffee, copper or fish. They must focus on manufacturing, financial and knowledge industries that cause them to go up the value chain with increasing margins, and which often produce monopolistic positions. Such companies can pay higher salaries for talent, and countries grow rich when the salaries go up rather than decline.

Another requirement for countries who want to progress is an educational policy that brings up the basic level of education in the nation, especially for girls since this will help prevent illiteracy, early marriages, early pregnancy and thus over-population. However, raising the educational level in a population will not help to increase its wealth unless there is also an industrial policy that produces opportunities for the educated. Without internal opportunities a boost in higher education is only likely to increase the propensity for the skilled and educated to emigrate.

Many factors can affect these basic principles. The point being stressed, however, is that if you want political power on the world stage you need economic power.

Culture, Country, City, Company, Person, Purpose, Passion, World can help you understand how to achieve this. It shows various Kondratieff waves dating all the way back to the 900s in China that usually caused sixty year prosperity waves. Kondratieff waves were due to all sorts of economic innovations such as printing, national markets, tax reform, maritime trade expansion, revolutionary raw materials, revolutionary production methods, revolutionary transportation methods, revolutionary communication methods and so on.

When we analyze Kondratieff waves and their impact on economic prosperity their benefits can often be viewed as the result of the *greater efficiencies they offered.* These efficiencies arose from new transportation modes that cut transportation costs and times (canals, railroads, roads, bridges); new trade routes along with the new products that then became available (spices, silk, dyes, etc.); cheaper or faster communication and information flows (woodblock printing, telecommunications); more efficient and inexpensive sources of energy (steam and electricity); cheaper market clearing mechanisms (fairs and trade markets and reduced tax systems); cheaper production techniques (steam motors, mass production), or any other "cheaper" improvements that added value.

Kondratieff waves are basically innovation waves. They are long-term waves of technological process which produce economic progress because they increase economic efficiencies by cutting costs, increasing productivity and by improving trade. Thus they boost economies and you should align with their modernizations so that your economy will flourish and you can

THE ART OF POLITICAL POWER

improve the standards of living. Sometimes they constitute grand political innovations that bring peace, improve national health through greater hygiene, improve general food availability or lower public taxes, but they always create large economic benefits that can make a country wealthy if it becomes an early adopter of the trends. With each new grand innovation wave the class structure of human society also changes in response to the new economic realities. If you are interested in how social movements develop according to Kondratieff waves or generational waves then you can turn to *Culture, Country, City, Company, Person, Purpose, Passion, World.*

Even if you are a country with virtually zero natural resources worth exploiting, as is the case of the small principality of Liechtenstein, you can still create a path for your country to become wealthy and grow in world significance. Liechtenstein insured its survival after the collapse of the Austro-Hungarian Empire, upon which it once depended, and set out on the road to prosperity by actually following the 29th lesson of the *Guanzi*.

Lesson 29 reads, "A strong state gains by winning the loyalty of small states but fails if it becomes too arrogant. A small state (Liechtenstein)gains by temporarily yielding to the big powers but fails if it breaks with its strong neighbors (first the Austro-Hungarian Empire and then Switzerland). Countries, whether large or small and strong or weak, each have their own particular circumstances and individual strategies for ensuring their preservation."

Liechtenstein was too small to survive on its own after the breakup of the Austro-Hungarian Empire, with whom it had a customs-union arrangement, so it later entered into an economic union with Switzerland. For Switzerland this was basically saving a small neighbor but it was also following the principles of lesson 29. Switzerland's aid stabilized a small neighbor but by doing so it secured a strategic buffer state and guaranteed that any debts that Liechtenstein owed it would be repaid. Liechtenstein cut its administrative costs due the arrangement, and was also able to base its economy on the stable Swiss franc.

Liechtenstein faced many barriers to development because it lacked sufficient money, technical knowhow and a large enough population to industrialize the same way as other European countries had done. Because it is a land-locked nation the lack of access to ocean transportation was a negative factor for trade and development. Studies show that land-locked nations experience 6% less economic growth than countries that not land-locked. Liechtenstein basically needed foreign investment in order to grow and develop so it cut taxes and instituted unprecedented pro-business policies as a recipe for economic growth. It basically attracted investment by offering freedoms and protections for business tax havens.

Today Liechtenstein has no external debts and one of the highest living standards in the world. Because of its small population Liechtenstein cannot do high quantity manufacturing so it specializes in highly specialized, high quality, and low volume manufacturing. It found its manufacturing niche.

When Singapore became independent of Malaysia in 1965 it also faced the economic challenge of building a nation out of a people with no national identity. Like Liechtenstein it had no natural resources so it faced the same difficult prospect of creating jobs through industrialization or modernization and pulling the population out of poverty to set them on a path to a better life.

To embark on his plans for modernization Singapore's leader, Lee Kuan Yew, chose the route of seizing authoritarian political power. This was a necessity to overcome its initial problem of potential social unrest and instability. Lee led a government solution to curtail individual freedoms, free speech and to institute authoritarian laws with strict enforcement so that peace and social stability would reign as the country began its shaky steps to modernization. This authoritarian rule was able to silence dissent and put a lid on potentially explosive underlying ethnic and social tensions that littered the state. Lee's emphasis on legalism curtailed violence and stabilized the nation. As long as population was promised prosperity and peace they put up with the temporary lack of political freedoms recognizing that these measures were to ensure social stability and were instituted for their own good.

As the social and political situation stabilized, Singapore embarked on a dramatic economic overhaul. It established policies that increased personal savings that enabled people to buy their own homes. Since it lacked resources and a manufacturing base, like Liechtenstein, the only way for Singapore to grow was by attracting foreign investment. Taxes and regulations were therefore kept low, companies were given tax incentives, and unions were banned to keep wages low in an attempt to attract foreign investors who had other opportunities to invest elsewhere throughout Southeast Asia.

The government also spent heavily on infrastructure projects to become a regional transportation hub just as Liechtenstein spent money on a hydroelectric dam for power generation after it created modernization plans. Singapore took drastic measures to cut out government corruption by raising government salaries, lowering the incentives for corruption, and increasing enforcement. To get away from cheap manufacturing and move up the value chain Lee Kuan Yew created technical schools, modern hospitals, and brought more specialized industries into Singapore. The government stressed an export-oriented economy and pushed for national productivity gains.

When starting out poor the problem for a country is attracting foreign investment (industrial technical know-how) because it doesn't have it sown capital base and the best means to do so is by educating citizens to the level where they are a desirable work force, by offering favorable tax treatments and good infrastructure, and by offering pro-business free market policies that entail special freedoms and protections for companies.

Across the world all human beings want to escape poverty, sickness and harm and experience pleasure and affluence instead. Most people really don't care about politics but just want to escape misery. They want to live a peaceful life with adequate living conditions. They want the chance to secure a gainful livelihood that lets them experience a degree of prosperity and peace of mind. As an aspiring leader, can you produce this for them? Marxism, in all its various forms, never does this. In China the common folk had a saying, "What do I care about the emperor and his doings? I till the fields and eat my fill. The emperor has nothing to do with me."

We can say most people just want to be left alone in their lives so that they can live in peace with little interference from the state. Of course they want adequate opportunities to achieve prosperity and an adequate freedom or scope for personal expression. They also want law and order in the state, access to a fair legal system of justice whenever necessary, and protection from invaders. But in general they just want to get along in their ordinary lives with very little interference from the state.

In short, most people want the ability to pursue a life of excellence with ample scope and opportunity but little intervention from the state. Political leaders using wise strategy are the only things that can make this possible for a populace, so that's the purpose behind the overall discussion on leadership and strategy. The purpose of leadership and strategy is to improve the lives of the people and provide them with pathways to a better life.

The famous sage Master Guigu (who wrote the *Guiguzi* that appeared in China towards the end of the Warring States period) once said that people naturally associate with one another (form coalitions) for several different purposes.

The first reason they form relationships with one another is due to common moral standards or they unify around common interest on an important issue. The second type of association involves the formation of communities, teams and parties. A third type of organization centers around profits and losses such as in the competitive sphere of business and trade. Families also serve as a basis of association, and the mentor-protégé relationship (such as the relationship between teachers and students) is also another reason people come together.

In the human world the very highest form of human associations are based on true emotional feelings or sincere relationships. We can call this associating principle "inner sincerity," or we can say that the highest form of association is based on deep human sentiments rather than some sort of dogma or ulterior purpose. This is the highest and purest form of relationship because it is a natural relationship that extends from heart to heart.

The next highest form of association is based on religious belief. Religions typically depend upon faith and reverence as their organizing principles. Each religion has its own dogma, but we can say that obedience to religious standards and disciplinary codes of conduct arises out of the principle of reverence.

Next, a third but even lower form of social association or relationship is that which is based purely on laws and regulations. This is using legalism as the bond or glue for a group of people, and any nation which tries to hold its society together based solely on legalism is already thrice removed from the highest form of social organization. A nation which emphasizes its legal system as its primary foundational basis will deteriorate over time because this entirely ignores the fundamental feelings of human nature that naturally exist between people that cause them to associate with one another, including their family bonds and friendship relationships.

The fourth form of organization is based solely on considerations of gain and loss, and relationships based on this principle typically use power, coercion and accounting to secure structure. Business, commerce and trade are examples of this type of relationship, and it's the lowest type of association between individuals because it is based purely on a materialistic monetary self-interest that can be counted. In fact, history shows that any form of organization based purely on monetary gains and losses (such as a cartel), without being closely bound to human relationships, can only endure for a short while.

Why should we bring up all these sorts of relationships? So that you remember that the world is a complicated place with many forms of competing relationships simultaneously operating in society. All these forms of association are always simultaneously in play crisscrossing one another and prompting human beings to act in this way or another.

You not only have to deal with all these various interrelationships to survive in the world, but you must learn to navigate them or harness them if you wish bring peace and prosperity to the world. That is the ultimate purpose of political power.

The rishis of ancient India spent a great deal of time pondering about how to manage society *so that there would be inner harmony among all the people*. A political leader should understand this goal. A leader who seeks the power to guide a nation must first recognize that there are all these conflicting, competing interests, and must develop a vision for guiding the people to a state of harmony *and* prosperity.

The American historian Rufus Fears said that a statesman is higher than a political leader and is distinguished by four special characteristics. He has a bedrock of principles that run his life, follows the dictates of his own moral compass, develops a vision to lead the people, and has the persuasive ability to build a consensus around his goals and vision. He convinces the people that his vision and goals are worth supporting and galvanizes them into implementing it.

With wisdom, persuasion and leadership skills you can persuade disparate groups to support your efforts that have an ultimate end of helping the people flourish in peaceful harmony with one another, and so that's the purpose of leadership, strategy and policy.

If you really want to be a great leader in the world you not only must cultivate wisdom, diplomacy and the ability to motivate others, but you must also understand all these factors that involve psychology, social structure, economics, politics and history. Then you can be really effective in helping to balance the world and cure it of its illnesses. The world is not owned by one person but by all the people of the world, and you become its leader by serving the people rather than by serving yourself.

In the various schools of spirituality across the world, masters are often saying that spiritual cultivation entails harmonizing the four elements (wind, water, fire and earth) of your physical body. The Chinese have extended this idea to the nation. Only if a nation's internal elements are balanced can it experience peace and prosperity in the world, which is why the Hindu sages were always pondering how to harmonize all the elements within the nation.

Political power can be achieved if you cultivate it through the methods of Master Guan. The question, once again, is to what purpose? The intended purpose of the text is a teaching guide for bringing about an auspicious state of peace, harmony and prosperity to a country. With this in mind, may you use Master Guan's teachings wisely.

SELECTED BIBLIOGRAPHY

Allison, Graham and Blackwill, Robert. *Lee Kuan Yew: The Grand Master's Insights on China, the United States, and the World*, The MIT Press, Massachusetts, 2013.

Bodri, Bill. *The I Ching Revealed*, Top Shape Publishing, Reno, Nevada, 2023.

Bodri, Bill. *Culture, Country, City, Company, Person, Purpose, Passion, World*, Top Shape Publishing, Reno, Nevada, 2018.

Cleary, Thomas. *Thunder in the Sky: On the Acquisition and Exercise of Power*, Shambhala Publications, Boston, Massachusetts, 1993.

Hedrick, Larry. *Xenophan's Cyrus the Great: The Arts of Leadership and War*, Truman Talley Books, New York, 2006.

Hu Jichuang. *A Concise History of Chinese Economic Thought*, Foreign Languages Press, Beijing, 1988.

Lau, D. C. *Confucius: The Analects (Lun yu)*, The Chinese University Press, Hong Kong, 1992.

Lau, D. C. *Mencius*, The Chinese University Press, Hong Kong, 1984. Lee, Walton. Wisdoms Way: 101 Tales of Chinese Wit, YMAA Publication Center, MA, 1997.

Legge, James. *The Chinese Classics (in 5 Volumes)*, SMC Publishing Inc., Taipei, 1994.

Li Fu Chen. *The Confucian Way*, KPI Ltd., London, 1987.

Low, C. C. & Associates. *Sun Zi's Art of War*, Canfonian Pte Ltd., Singapore, 1993.

Maverick, Lewis, T'an Po-fu and Wen Kung-wen. *Economic Dialogues in Ancient China: Selections from the Kuan-tzu*, Carbondale, Illinois, 1954.

Parker, Edward H. "Prussian Kultur 2,500 Years Ago," *Asiatic Review*, Vol. 14, No. 39, July, 1918.

Parker, Edward H. "Kwan-Tsz," *The New China Review*, Vol. III, No. 1, February, 1921.

Parker, Edward H. "Kwan-Tsz," *The New China Review*, Vol. III, No. 6, December, 1921.

Rickett, W. Allyn. *Kuan-Tzu*, Hong Kong University Press, Hong Kong, 1965.

Rickett, W. Allyn. Guan-Zi: *Political, Economic, and Philosophical Essays from Early China (A Study and Translation)*, Princeton University Press, Princeton, New Jersey, 1985.

Sawyer, Ralph D. *The Seven Military Classics of Ancient China*, Westview Press, Boulder, Colorado, 1993.

Sawyer, Ralph D. *The Six Secret Teachings on the Way of Strategy*, Shambhala, Boston, Massachusetts, 1996.

Ssu-ma Ch'ien. *The Grand Scribe's Records (Volume VII): The Memoirs of Pre-Han China* (edited by William Nienhauser), SMC Publishing Inc., Taipei, 1994.

Sun Haichen. *The Wiles of War: 36 Military Strategies from Ancient China*, Foreign Languages Press, Beijing, 1991.

Wang Congren. *Tales about Prime Ministers in Chinese History*, Hai Feng Publishing Co., Hong Kong, 1994.

Watson, Burton. *The Tso Chuan*, Columbia University Press, New York, 1989.

Zhang Liqing and Jiang Tao. *A Collection of Guan Tzu's Sayings*, Qi Lu Press, 1992.

ABOUT THE AUTHOR

Bill Bodri is the author of several investment, marketing, health and peak performance books including the following:

Breakthrough Strategies of Wall Street Traders
Super Investing
Hard Yield Investments, Hard Assets and Asset Protection Strategies
Bankism

Culture, Country, City, Company, Person, Purpose, Passion, World
The Art of Political Power
The I Ching Revealed

How to Create a Million Dollar Unique Selling Proposition
The Claude Hopkins Rare Ad Collection
Move Forward
Quick, Fast, Done

Visualization Power
Sport Visualization

Detox Cleanse Your Body Quickly and Completely
Look Younger, Live Longer
Super Cancer Fighters
Neijia Yoga

The four books most relevant to this one would be *Culture, Country, City, Company, Person, Purpose, Passion, World* (highly recommended), *Bankism*, *Super Investing* and *Quick, Fast, Done*.

Made in the USA
Columbia, SC
04 June 2024

36639891R00089